Inorganic and Analytical Chemistry Laboratory Manual

无机及分析化学实验(双语)

浙江大学化学系

主　编　曾秀琼

副主编　蔡吉清　何桂金　徐孝菲

科 学 出 版 社

北 京

内 容 简 介

　　本书是为适应信息化和国际化教育需求而编写的。全书共 5 章，包括绪论、基本操作和常用仪器、基础无机化学实验、基础分析化学实验、综合设计实验。每个实验配有详尽的背景知识、实验原理、实验步骤、安全提示和废弃物处理等；绝大部分实验有拓展部分，以满足教学高阶性的需求。本书双语模式独具特色，以满足不同英文水平学习者的需求。本书还配套了 62 个最新完成的高质量视频，读者扫描书中二维码即可观看学习。

　　本书可作为高等学校化学、化工、医学、药学、农学、生命科学、材料科学和环境科学等专业本科生的实验教材或教学参考书，也可供相关人员学习或参考阅读。

图书在版编目（CIP）数据

无机及分析化学实验 = Inorganic and Analytical Chemistry
Laboratory Manual：双语版：汉文、英文 / 曾秀琼主编. -- 北京：
科学出版社，2025. 1. --ISBN 978-7-03-080234-7

Ⅰ. O61-33；O652.1

中国国家版本馆 CIP 数据核字第 2024JS1668 号

责任编辑：丁　里 / 责任校对：杨　赛
责任印制：吴兆东 / 封面设计：迷底书装

科 学 出 版 社 出版
北京东黄城根北街 16 号
邮政编码：100717
http://www.sciencep.com

北京华宇信诺印刷有限公司印刷
科学出版社发行　各地新华书店经销
*
2025 年 1 月第 一 版　开本：787×1092　1/16
2025 年 3 月第二次印刷　印张：10 3/4
字数：279 000

定价：49.00 元
（如有印装质量问题，我社负责调换）

Preface

The Inorganic and Analytical Chemistry Laboratory Course has been established over twenty years at Zhejiang University (ZJU). The course team places emphasis on teaching research and practice, has won numerous teaching awards, such as Provincial First-class Undergraduate Courses in 2020, the Design Star at National Teaching Design Innovation Contest in 2021, the First Place Prize Teaching Achievement Awards of ZJU in 2021, the Ideological and Political Theory Demonstration Courses for Undergraduates of ZJU in 2023, Zhejiang Province's "Internet + Teaching" Model Classroom in 2019, etc.

To promote international teaching, the course team completed writing a bilingual laboratory manual in the fall of 2021. Over the course of six semesters, the team applied the manual in their courses. The manual's bilingual layout evolved from an initial design of English on the left page with full Chinese translation on the right, finally to a layout with full English in a left column and some Chinese notes in a right column. To accomodate students with varying English proficiency, the manual also includes Chinese translation of long phrases, entire sentences, and even whole paragraphs.

To promote blended online and offline teaching, the course team spent over two years developing a bilingual MOOC, which was launched on China large-scale education platform. There are sixty-two videos embedded in the manual, sixty of which are derived from this bilingual MOOC.

The aims of this manual are: (1) establishing the scientific notions of laboratory safety and environmental protection, (2) cultivating scientific literacy that is standardized, rigorous, and seeking truth from facts, (3) building the scientific spirits of innovation and team-work. The manual covers the fundamental principles, knowledge, and techniques in inorganic chemistry experimentation and analytical chemistry experimentation. It includes the following chapters: introduction, basic operations and common instruments, fundamental inorganic chemistry experiments, fundamental analytical chemistry experiments, comprehensive and designing experiments.

To respond to the idea that 'green mountains are gold mountains', this manual implements the '3R' concept of green chemistry to reduce reagent consumption and laboratory waste. The ongoing macro analytical chemistry experiments are designed with a relative error less than or equal to 0.2%, a titration volume of 15-20 mL, and a weighing mass less than or equal to 0.1 g. Additionally, please notice the following two statements: (1) Unless otherwise specified, all experimental water used is deionized water. (2) If there are discrepancies between the expressions, reaction conditions, or reagent quantities in this manual and those shown in the embedded bilingual MOOC videos, the information in this manual shall take precedence.

This manual was written by Xiuqiong Zeng, Jiqing Cai, Guijin He, and Xiaofei Xu. Each writer is indicated in the corresponding parts. Xiuqiong Zeng is responsible for the overall content of the manual, Stacy Zeng and Zexi Mao for editing English grammar and expressions, and Jiqing Cai for some illustrations.

In the construction of this course and the bilingual manual, we obtained strong support from the Undergraduate School of ZJU, the Department of Chemistry of ZJU, the National Demonstration Center for Chemistry Experimental Teaching of ZJU, and the course team members, especially Professor Min Wang. Throughout the compiling this book, numerous relevant teaching materials have been consulted. We would like to express our heartfelt thanks.

Due to the limited level and experience of the writers, it is inevitable that there will be some negligence and deficiencies in this manual, and we sincerely thank readers for any criticism.

Writers

August, 2024, in Hangzhou, Zhejiang, China

前　言

  浙江大学的无机及分析化学类实验课程已开设二十多年。课程团队致力于教学研究与实践，取得多项教学成果，如浙江省本科高校 2020 年度省级一流课程、2021 年全国高校混合式教学设计创新大赛设计之星、2021 年浙江大学教学成果奖一等奖、2023 年浙江大学校级本科课程思政示范课程、2019 年浙江省"互联网+教学"示范课堂等。

  为了推动国际化教学，课程团队于 2021 年秋编写完成双语讲义，在校内使用了 6 个教学周期。双语模式由最初的左页英文-右页中文，最后到左栏英文-右栏中文注释，并增加了长词组、整句甚至整段的中文注释，以满足不同英文水平学生的学习需求。

  为了推动线上线下混合式教学，课程团队历时两年多建成一门双语慕课，并在国内大型教育平台上线。本书配套了 62 个视频，其中 60 个来自该双语慕课。

  编写本书的目的是：(1)树立实验安全和绿色环保的科学理念，(2)塑造规范严谨和实事求是的科学素养，(3)培养勇于创新和团结互助的科学精神。本书内容涵盖无机化学实验和分析化学实验的基本理论、基本知识和基本技能，包括以下章节：绪论、基本操作和常用仪器、基础无机化学实验、基础分析化学实验、综合设计实验等。

  为了响应"绿水青山就是金山银山"，本书将绿色化学"3R"理念落到实处，降低试剂消耗量和实验室废液产生量。本书以相对误差 ≤ 0.2%、"滴定体积为 15～20 mL，且称量质量 ≥ 0.1 g"来设计常量分析化学实验。此外，请关注以下两点说明：(1)除特别指出外，实验用水全部为去离子水。(2)若本书中的个别表达、反应条件或试剂用量等与配套的双语慕课视频中的不一致，以本书为准。

  参加本书编写工作的教师有曾秀琼、蔡吉清、何桂金和徐孝菲，具体编者在书中相应部分均有注明。曾秀琼负责全书统稿，Stacy Zeng 和毛泽茜负责英文语法和表达的校对，蔡吉清负责部分插图。

  在本课程和双语教材的建设中，得到了浙江大学本科生院、浙江大学化学系、浙江大学国家级化学实验教学示范中心等，以及本课程团队成员，特别是王敏教授的大力支持。在本书编写过程中参考了很多国内外相关教学资料。在此，一并表示衷心的感谢。

  由于编者水平和经验有限，书中难免有疏漏和不足之处，恳请广大读者批评指正。

编　者

2024 年 8 月于杭州

目　录

Preface
前言
Chapter 1　Introduction 绪论 ···················· 1
  1.1　Chemistry Laboratory Safety Policies 化学实验室安全制度 ···················· 1
  1.2　Requirements for the Course 课程要求 ···················· 6
  1.3　Guidelines for Lab Notebook, Pre-lab, and Post-lab Report
     实验记录本、预习报告和实验报告指南 ···················· 9
  1.4　Introduction to Titrimetric Analysis 滴定分析法概述 ···················· 12
  1.5　Significant Figures and Data Treatment 有效数字和数据处理 ···················· 16
Chapter 2　Basic Operations and Common Instruments 基本操作和常用仪器 ···················· 22
  2.1　Cleaning and Drying Glassware 玻璃器皿的洗涤和干燥 ···················· 22
  2.2　Transferring Reagents 试剂的取用 ···················· 24
  2.3　Use of Volumetric Glassware 量器类玻璃器皿的使用 ···················· 25
  2.4　Separation Techniques 分离技术 ···················· 32
  2.5　Common Instruments 常用仪器 ···················· 36
Chapter 3　Fundamental Inorganic Chemistry Experiments 基础无机化学实验 ···················· 47
  Expt. 1　Separation and Identification of Common Cations 常见阳离子的分离和鉴定 ···················· 47
  Expt. 2　Separation and Identification of Common Anions 常见阴离子的分离和鉴定 ···················· 53
  Expt. 3　Determination of Reaction Rate Constant, Activation Energy and Influencing Factors
     反应速率常数、活化能及影响因素的测定 ···················· 60
  Expt. 4　Preparation of Alum and Large Crystal Cultivation 明矾的制备及其大晶体培养 ···················· 66
Chapter 4　Fundamental Analytical Chemistry Experiments 基础分析化学实验 ···················· 72
  Expt. 5　Preparation and Standardization of the Solutions for Acid-base Titration
     酸碱滴定溶液的配制与标定 ···················· 72
  Expt. 6　Preparation and Standardization of the Solutions for Redox Titration
     氧化还原滴定溶液的配制与标定 ···················· 77
  Expt. 7　Preparation and Standardization of the Solution for Complexometric Titration
     配位滴定溶液的配制与标定 ···················· 84
  Expt. 8　Preparation and Standardization of the Solution for Precipitation Titration
     沉淀滴定溶液的配制与标定 ···················· 88
  Expt. 9　Practise of Basic Operations for Titrimetric Analysis 滴定分析基本操作训练 ···················· 92
  Expt. 10　Determination of the Molecular Weight of an Unknown Diprotic Organic Acid
     未知二元有机酸分子量的测定 ···················· 95
Chapter 5　Comprehensive and Designing Experiments 综合设计实验 ···················· 98
  Expt. 11　Determination of the Fluoride Content in Tea Samples by FISE
     氟离子选择电极法测定茶叶中氟离子含量 ···················· 98

Expt. 12   Preparation and Characterization of Mohr's Salt 莫尔盐的制备及表征 ………… 104

Expt. 13   Preparation and Characterization of Copper Methionine
蛋氨酸铜的制备及表征 ……………………………………………………… 107

Expt. 14   Preparation and Photosensitivity Test of Potassium Trioxalatoferrate(III) Trihydrate
三水合三草酸合铁(III)酸钾的制备及光敏性测试 ……………………… 111

Expt. 15   Determination of the Charge Number of Trioxalatoferrate(III) Complex Ion
三草酸合铁(III)配离子电荷数的测定 ……………………………………… 115

Expt. 16   Spectrophotometric Determination of Iron Content 分光光度法测定铁含量 ····· 118

Expt. 17   Synthesis and Characterization of Three Kinds of Cobalt-ammine Coordination Compounds
三种钴氨配合物的制备及表征 ……………………………………………… 126

Expt. 18   Determination of Cement Composition 水泥组分测定 ……………………… 132

Expt. 19   Preparation and Determination of Three Kinds of Sodium Phosphate Hydrates
三种磷酸钠盐水合物的制备及测定 ………………………………………… 139

Expt. 20   Preparation of Heteropolyacid Salts and Determination of Conversion Kinetics
杂多酸盐的制备及其转化动力学测定 ……………………………………… 144

APPENDIX ……………………………………………………………………………… 151

Appx. 1   Illustration of Common Glassware and Apparatus 常用玻璃器皿和仪器插图 ···· 151

Appx. 2   Vocabulary of Common Professional Words and Phrases
常用专业词汇中英文对照表 ………………………………………………… 152

Appx. 3   Density, Content and Concentration of Common Concentrated Acids and Bases
常用浓酸、浓碱的密度、含量和浓度 ……………………………………… 157

Appx. 4   Common Indicators 常用指示剂 …………………………………………… 157

Appx. 5   Color of Common Ions and Inorganic Compounds
常见离子和无机化合物的颜色 ……………………………………………… 159

Appx. 6   Bilingual Videos Embedded in the Manual 本书插入双语视频清单 …………… 160

Appx. 7   Atomic Weight of Elements 元素的原子量 ………………………………… 162

Chapter 1　Introduction
绪　　论

This is a laboratory-based course, and a chemistry laboratory can be a dangerous place. Safety and health are the top priorities in this course. Be aware of the dangers and pay special attention to your surroundings at all times. Before starting the first experiment, review the following safety policies, dress code, course requirements[1] and other policies and regulations. Failure to comply with these policies and regulations will result in dismissal from the laboratory[2], and even removal from the course.

[1] 安全制度，着装规范，课程要求

[2] 被赶出实验室

1.1　Chemistry Laboratory Safety Policies
化学实验室安全制度

1.1.1　General Safety Policies

(1) Wear a lab coat and safety goggles at all times[3] in the laboratory, not just when an experiment is in progress. Avoid skin contact with all chemicals, and wear protective gloves[4] whenever needed (Fig. 1-1). Refer to "1.1.2　Dress Code" for more details.

[3] 全程穿实验服和戴护目镜

[4] 防护手套

lab coat　　　safety goggles　　　protective gloves

Fig. 1-1　Personal protective equipment (PPE)

(2) Arms, legs, and feet must be completely covered[5] with appropriate clothing and shoes. Refer to "1.1.2　Dress Code" for more details.

[5] 胳膊、腿部和脚部需被全部遮住

(3) Long hair must be tied back[6] to reduce possible damages. **Note:** Long hair is defined as hair at a length that can be tied back.

[6] 长发必须扎起来

(4) Never wear contact lenses in the laboratory, as they can trap corrosive or volatile materials[7] which may damage the eyes. Furthermore, they may interfere with eye washing and treatment in case of an emergency.

[7] 因为腐蚀性和挥发性物质会陷入隐形眼镜内

[8] 洗眼器、紧急喷淋器、灭火器、灭火毯、消防沙、急救箱和化学品泄漏处理包

[9] 重复鸣笛和火警信号灯闪烁警示从大楼撤离

[10] 用手将气体扇向鼻子并小心吸入

[11] 绝不尝试未经许可或安排的实验

[12] 化学品安全说明书

[13] 危险化学品安全须知卡

[14] 不可将个人移液管或滴管伸入公用液体试剂容器中

[15] 有缺损、破裂或损坏的玻璃器皿

[16] 绝不加热密封体系，如盖紧的碘量瓶

[17] 沸石

[18] 隔热手套

[19] 涉及有毒有害气体的操作在通风橱内进行

[20] 易燃材料

(5) Know the exact location of and how to use each safety equipment in the laboratory, such as the eyewash fountain, emergency (safety) shower, fire extinguisher, fire blanket, fire sand, first aid kit, and chemical spill kit[8]. However, do not use the emergency showers for non-emergencies, as they are designed to deliver water at a very high flow rate.

(6) Be familiar with all of the exits in the laboratory. A repeating siren and flashing of the FIRE indicator are the building's evacuation signals[9]. If this alarm goes off, turn off any open flames and leave the building as quickly as possible.

(7) NEVER eat, drink, chew, or smoke in the laboratory. Never taste any chemicals. Never smell any chemicals without permission. If being instructed to smell a chemical, waft the vapors toward your nose with your hand and inhale cautiously[10].

(8) Never perform experiments alone in the laboratory. Never attempt any unauthorized or unassigned experiments[11]. Explicitly follow the experimental procedures, check and double check the identity of all materials before use.

(9) Be familiar with the Material Safety Data Sheets (MSDS)[12] or Hazardous Chemicals Notification Cards[13] before working with any chemicals. These sheets (or cards) contain information about the physical properties of each chemical and identify any hazards associated with them. They also provide details on special handling precautions and the protective equipment required when working with each chemical.

(10) NEVER pour unused or excess chemicals back into their original stock containers. Share them with others or dispose of them in the designated container. Never insert a personal pipette or dropper into a shared liquid reagent container[14]. Close the cap and lid tightly after each use.

(11) Keep glassware away from the edge of the bench. Inspect glassware before and after each use. Discard or repair any cracked, broken, or damaged glassware[15]. Do not pick up broken glass with unprotected hands.

(12) NEVER heat a "closed system", such as a stoppered iodine flask[16]. When heating liquids directly, always add boiling chips[17] so that the liquid boils at a safe rate. Always use heat-resistant gloves[18] to handle hot materials or apparatus.

(13) All operations involving noxious or poisonous gases or vapors must be carried out in the fume hood[19]. Note: **NEVER put your head inside the hood.**

(14) When using a flame, ensure that no flammable materials[20] are nearby. When heating or carrying out reactions in a test tube, never point

its mouth at anyone[21], as the contents can boil over quickly and splash.

(15) When cutting glass tubing or inserting it into stoppers[22], protect your hands by using a towel or cotton-gloves. Glass tubing should be lubricated with glycerin[23], soapy water, or water to aid the insertion.

(16) To dilute a concentrated acid, always pour it slowly into water, while stirring the solution[24]. Never pour water into concentrated acids, otherwise, local heating and density effects may cause the solution to splash.

(17) Handle chemical and water spills immediately[25]. Neutralize acid or base spills before cleaning them up. Boric acid (H_3BO_3) solution is available to neutralize alkali spills[26], and sodium carbonate (Na_2CO_3) powder is provided to neutralize acid spills. If the spill involves dangerous chemicals, inform the instructor or TA immediately.

(18) Dispose of waste properly, especially broken glass. Only nonhazardous, water-soluble materials can be poured down the sink[27]. Never pour solid disposals or untreated hazardous solutions down the sink. Never discard chemicals or chemical-contaminated paper in the trash cans.

(19) NEVER leave heated apparatus unattended[28]. Turn them off or inform the instructor or your partner if you have to leave.

(20) All items in the laboratory, such as instruments and chemicals, should not be taken out without permission.

(21) Report all accidents to your instructor or TA immediately.

(22) Keep your working area clean and in order.

1.1.2　Dress Code

Students must not have any bare skin exposed on their arms or feet. Read the following dress code carefully.

1. Lab Coats

(1) Lab coats must fit properly, covering the arms to the wrists without excess length and the legs to at least below the knees[29].

(2) Lab coats must be **fully buttoned**. Open lab coats will not be permitted at any time.

2. Pants

(1) Pants must cover the entire legs, ending at the top of the shoes[30].

(2) Pants must cover the ankles, including the socks. Exposed ankles or socks will not be permitted.

(3) Leggings, exercise pants, and other tight-fitting synthetic materials are **NOT** permitted[31] because they can accelerate the transfer of chemicals to the skin and are difficult to remove in case of an emergency.

[21] 绝不将试管口朝向任何人

[22] 将玻璃管插入塞子里

[23] 用甘油润滑

[24] 稀释浓酸时，需边搅拌边将其缓慢倒入水中

[25] 立即处理溅出的化学品或水

[26] 硼酸溶液可用来中和碱性溅出物

[27] 只有无危害的水溶性物质可倒入水槽

[28] 绝不放任加热装置无人看管

[29] 实验服需遮住整条胳膊且不超出手腕，下端至少过膝盖

[30] 裤子需遮住整个腿部，下端直至鞋子上方

[31] 严禁穿打底裤、健身裤和其他合成材质的紧身裤

[32] 鞋子需盖住整个脚面，且不能露出袜子

[33] 严禁穿凉鞋、拖鞋、露脚趾鞋、露后跟鞋和阔口鞋

[34] 严禁将靴子(包括雨靴)穿在裤子外面

[35] 一定要穿袜子，作为多一层保护

[36] 异物进眼睛

[37] 用洗眼器的流水冲洗受伤的眼睛

[38] 稀释和冲洗掉化学品是非常重要的

[39] 难溶性的异物

[40] 不能揉眼睛

[41] 棉签

[42] 眨眼有助于产生更多眼泪而利于异物排出

3. Shoes

(1) Shoes must completely cover the feet, with no exposed socks[32].

(2) Sandals, slippers, open-toe shoes, open-back shoes, and boat shoes are **NOT** permitted[33] at any time.

(3) Boots (including rain boots) that go over the pants are also **NOT** permitted[34], as they leave a space between the pant and the boot. This is dangerous because chemicals can easily spill into the boot.

4. Socks

Socks must be worn with ALL shoes as an extra layer of protection[35].

1.1.3　Some Emergency First Aid

It would be untruthful to say that there is no element of risk in a chemistry laboratory; therefore, emergency first aid is of utmost importance. **Note:** The following is only the first aid available in the laboratory. Report all the accidents to the instructor or TA, and send the severely injured person to the hospital immediately.

1. Foreign Objects in the Eye[36]

Ⅰ. Chemicals in the Eye

If chemicals are splashed into your eye(s), the affected eye(s) must be flushed with a continuous stream of water from an eyewash fountain[37] for at least 15 minutes. **Remember:** The longer a chemical remains in the eye, the more damage it can cause. Diluting and washing the chemical away are extremely important[38], opening the eyelids as wide as possible.

Ⅱ. Other Foreign Objects in the Eye

When insoluble foreign objects[39] get into the eye, especially sharp materials like metal or glass, remember the following suggestions:

(1) **Do not rub the eye**[40], as this can cause a scratch on the surface of the eyeball.

(2) **Do not touch the eye with anything**. Fingers, cotton swabs[41], and other objects will not help remove foreign objects and may cause serious damage to the eye.

(3) **Do blink**. Blinking will help produce more tears and aid in removing foreign objects[42]. **Note:** Do not blink when sharp materials get into the eye.

2. Fire

Ⅰ. Work Place on Fire

The method used to extinguish a fire depends on the size of the fire and the substances that are burning.

(1) If only drops of flammable liquid have been accidentally ignited and no other flammable liquid is nearby, the fire can be extinguished by covering it with fire sand or a wet towel[43].

(2) If a large chemical fire occurs or electrical equipment catches fire, carbon dioxide (CO_2) fire extinguishers should be used. **Note:** Carbon dioxide fire extinguishers cannot be used on fires involving most reactive metals[44], such as Mg, Na, because CO_2 can react vigorously with the burning metal and make the fire worse.

(3) Foam extinguishers[45] are used on fires involving other flammable solids and liquids.

(4) If the fire cannot be extinguished immediately, evacuate the laboratory and call the fire department.

Ⅱ. Clothing or Hair on Fire

If possible, remove the burning clothing and use the emergency shower, fire blanket, or other heavy clothing immediately. If no emergency shower or fire blanket available, the common advice is to "stop, drop, and roll"[46] (Fig. 1-2). Stop running, drop to the ground, and roll from side to side to extinguish the fire. **Note:** Moving around will only worsen the fire.

Fig. 1-2   Stop, drop, roll (from left to right)

3. Chemical Burns

When handling with the chemicals, caution must be taken to prevent contact with the skin. In particular, most acids, alkalis, oxidizing and reducing chemicals are corrosive to the skin[47]. When handling these chemicals, keep your hands away from your eyes and face until they have been thoroughly washed. A chemical's corrosiveness is usually proportional to its concentration[48], so wash your hands regularly after using any chemicals. If any chemicals is spilled on the skin, wash immediately with plenty of water. **Suggestion:** Wear gloves when handling corrosive chemicals.

In the case of a major chemical spill, in which a large portion of your body or clothing is affected, use the emergency shower immediately[49], and remove any contaminated clothing for thorough washing.

[43] 通过覆盖消防沙或湿毛巾来灭火

[44] 二氧化碳灭火器不能用于绝大部分活泼金属引起的火灾

[45] 泡沫灭火器

[46] 停下、蹲下、打滚

[47] 大部分酸、碱、氧化性和还原性化学品会腐蚀皮肤

[48] 化学品的腐蚀性常与其浓度成正比

[49] (大面积化学品泄漏时)立即使用紧急喷淋装置

4. Cuts

When minor cuts occur, wash your hands and the affected area with clean water and antibacterial or mild soap[50], then wrap with a sterile bandage[51].

When deep cuts occur, it is crucial to stop the bleeding as quickly as possible[52]. Wrap the affected area with a towel or bandage, then continue applying pressure to stop the blood loss and reduce the risk of infection.

5. Thermal Burns

Rinse the affected area immediately with cool water for at least 15 minutes. This may limit the extent and severity of the burn. Then, apply an antibiotic ointment to aid in healing and reduce the risk of infection[53]. **Note:** Do not use ice, this may worsen the injury to the skin.

6. Electrical Shock

Never rush into the area where an accident occurred or touch the victim. Wear protective gloves or use a non-conducting material[54] to cut off the switch. Then, move the victim to an open area and start resuscitation[55] if needed.

扫一扫　视频 1-1　化学实验室安全

1.2　Requirements for the Course
课 程 要 求

1.2.1　Basic Requirements

Students who attend this course must comply with all laboratory safety policies, rules, and the following requirements.

(1) At the beginning of the course, complete the laboratory safety education and training, pass the Safety Examination[56], and sign the Safety Acknowledgement Sheet[57]. **Note:** If the student does not pass the Safety Examination, he/she will not be allowed to attend the course.

(2) Each student will be assigned a laboratory locker[58] with a set of glassware and other apparatus. Check each item and replace any chipped, cracked, or other imperfect glassware[59]. Imperfect glassware is a major safety hazard and must be discarded. **Note:** Never store any chemicals in the lab locker without permission.

[50] 抗菌皂或温和皂
[51] 包上创可贴

[52] 当创口很深时，尽快止血是关键

[53] (烧伤时)涂抹抗生素软膏，有助愈合和减小感染风险

[54] (触电时需用)绝缘材料
[55] 心肺复苏

[56] 通过安全考试
[57] 签署安全知情书

[58] 实验储物柜

[59] 缺口、裂痕或其他有缺陷的玻璃器皿

(3) Always bring safety goggles, lab coat, laboratory manual[60], notebook, pre-lab and post-lab reports to the laboratory. Keep coats, backpacks, and other non-essential materials away from the working area. **Note: NEVER** sit or lean on the lab benches, as chemicals may be present, even if they are not visible to the naked eyes.

[60] 护目镜、实验服、实验教材

(4) Before the lab class, get familiar with the experiment's objectives, principles, operations, procedures and precautions[61]. To achieve these goals, first carefully watch the online teaching videos, thoroughly read the laboratory manual, then complete all the pre-lab assignments. **Note:** There is always an online quiz for each experiment to assess the students' pre-lab activities.

[61] 熟悉实验目的、原理、操作、步骤和注意事项

(5) Arrive at the laboratory ahead of time, sign in upon arrival, and ensure everything is in order and prepared. Check glassware before use, as small cracks can cause it to break or explode.

(6) During the lecture given by the instructor, take notes of important details and special instructions[62].

[62] 听讲解时，记录重点细节和特别说明

(7) During the experiment, keep quiet, perform the experiment carefully, observe conscientiously[63], record all the data and observations truthfully in the notebook, and keep the working area clean and organized.

[63] 认真完成实验，细致观察

(8) To avoid contamination of public supplies, do not use personal spoons or dropper to transfer the shared chemicals.

(9) Never operate an unknown instrument without permission. Make a record in the registration form after using any valuable instrument[64].

[64] 使用任何贵重仪器后需登记

(10) There are several different labeled waste containers[65] in the laboratory. Dispose of all waste in the suitable containers. The broken glass container is for broken glass **ONLY**. The solid waste container is for chemicals, gloves, or other contaminated materials used in the laboratory. The liquid waste container is for chemical solutions. The trash cans are for regular lab disposal. Do not fill the containers beyond 90% capacity.

[65] 贴有标签的废弃物容器

(11) After finishing the experiment, turn off all equipment, clean all apparatus, ensure they are properly stored, and then clean your working area.

(12) Have your notebook checked and signed by the instructor or TA.

(13) Wash hands thoroughly, sign out before leaving the laboratory.

(14) Students on duty clean the laboratory, empty the trash cans[66], close the windows, and turn off the faucets and electrical switches.

[66] 值日生打扫实验室，清空垃圾桶

(15) After the lab class, complete the post-lab report then hand it in the next class.

1.2.2 Attendance Policy

Attendance is very important for a lab course.

(1) Anybody who arrives 15 or more minutes late will not be allowed to enter the laboratory.

(2) Anyone who misses class more than three times will fail the course. Valid excuses for missing class include a signed or stamped sick note from an official hospital[67], a family emergency, or a participation in an important university activity.

[67] 有效请假条包括正规医院签字或盖章的病假条

(3) Even with a permitted excuse, you are still required to complete a make-up lab[68] and accomplish all the assignments (online pre-lab quiz, pre-lab report and post-lab report, etc), otherwise no point will be awarded for that lab.

[68] 即使获准请假，仍需补做实验

1.2.3 Grading for this Course

Grading for this course includes the on-line pre-lab quiz[69] (10%), online activities[70] (10%), final written examination (15%), final operation examination[71] (15%), general performance (35%, including but not limited to, pre-lab report, post-lab report, research report, class activities, attendance, overall completeness of tasks, cleanliness and orderliness, time management), and others (15%, teamwork, equipment intactness[72], lab notebook, on-duty, etc).

[69] 线上预习考试
[70] 线上学习情况
[71] 期末操作考试

[72] 仪器完好

Grading of the reports includes neatness, quality of writing, accuracy of data and results, discussion, analysis[73], and other relevant factors.

[73] 书写工整，写作质量，数据及结果的准确，讨论，分析

There may be some changes in grading from semester to semester, and students will be informed at the beginning of the semester.

1.2.4 Safety Acknowledgement Sheet

You must sign the Safety Acknowledgement Sheet before you are allowed to work in the lab. If you have any questions about these rules and procedures, please ask your instructor or TA.

I, the undersigned student, have received safety training, understand it and agree to comply with the safety policies. I understand the importance of protection in the lab. I will wear safety goggles and a lab coat at all times in the lab. I have been warned of the dangers of wearing contact lenses[74] in the lab and will not wear them in the lab. I know the exact location of and how to use each safety equipment in the lab.

[74] 隐形眼镜

I do know that I will be immediately dismissed from the lab if I fail to comply with the safety policies of the chemistry laboratory.

Name: _____ Student ID: _____ Signature: _____ Date:_____

扫一扫　视频 1-2　课程要求

1.3　Guidelines for Lab Notebook, Pre-lab, and Post-lab Report
实验记录本、预习报告和实验报告指南

1.3.1　Guidelines for Lab Notebook

In all research and industrial work, it is essential to keep good records for future reference; otherwise, work may be misinterpreted, lost, or unnecessarily repeated. **Note:** NEVER record any original data or observations on scraps of paper[75] or on the lab manual.

[75] 纸片

Here are some guidelines to be followed.

(1) Have a BOUND lab notebook with pre-recorded consecutive page numbers[76]. A spiral or loose-leaf[77] notebook is not allowed.

[76] 连续标好页码并装订成册的实验记录本

[77] 螺旋或活页的

(2) Use a ballpoint pen or gel ink pen for all entries to ensure that marks will not smear nor be erasable[78]. **Note: Never use a pencil for entries.**

[78] 记录不会被弄脏或被擦去

(3) Write your name, student ID, telephone number, and other information on the outside front cover of the notebook.

(4) Prepare your notebook each week before coming to the laboratory.

(5) Always start a new page[79] for each new experiment.

[79] 另取一页

(6) Always record data directly at the time it is collected or observations are obtained.

(7) **NEVER** tear a page out[80] of the notebook or erase a laboratory record. If a mistake is made in the notebook, do not cross it out completely, just put a single line through it so that it can still be read. Then write the correction nearby[81].

[80] 撕页

[81] 只需在错误处画条线使其仍可被读。然后在附近更正

(8) **NEVER** change or make up (fabricate)[82] data or observations; otherwise 0 will be awarded for the lab, and it may even result in a failing grade for the course.

[82] 伪造

(9) Here are some items that should be written in the notebook.

1) Title of the experiment and date. Always write this at the top.

2) Tables. They will be used to record all data and observations during the experiment. Always create your own tables prior to the experiment if no table is provided, reproduce the tables from the lab manual if they are available[83].

[83] 若无现成表格，需事先制作好。若有现成表格，需复制

There are six typical tables for the chemistry lab course. Table 1-1, Table 1-3 and Table 1-5 are created on the lab notebook before coming to the lab, filling them during the experiment. Table 1-2, Table 1-4 and

Table 1-6 are created on the post-lab reports, calculating the data and evaluating the results[84].

[84] 课前在记录本上制作好表格，课中填入，课后完成数据计算和结果分析
[85] 常见阳离子鉴定

Table 1-1   Identification of common cations[85] (on lab notebook)

Items	Procedures	Observations
test of Pb^{2+}	5 drops of 0.1 mol·L^{-1} Pb(NO$_3$)$_2$ + 1 drop of 6 mol·L^{-1} HAc + 5 drops of 0.1 mol·L^{-1} K$_2$CrO$_4$	
test of Cu^{2+}	3 drops of 0.1 mol·L^{-1} CuSO$_4$ + 1 drop of 2 mol·L^{-1} HAc + 1 drop of 0.1 mol·L^{-1} K$_4$[Fe(CN)$_6$]	

Table 1-2   Identification of some common cations (on post-report)

Items	Procedures	Observations	Expressions
test of Pb^{2+}	5 drops of 0.1 mol·L^{-1} Pb(NO$_3$)$_2$ + 1 drop of 6 mol·L^{-1} HAc + 5 drops of 0.1 mol·L^{-1} K$_2$CrO$_4$	yellow precipitates	PbCrO$_4$ forms Pb^{2+} + CrO$_4^{2-}$ === PbCrO$_4$↓
test of Cu^{2+}	3 drops of 0.1 mol·L^{-1} CuSO$_4$ + 1 drop of 2 mol·L^{-1} HAc + 1 drop of 0.1 mol·L^{-1} K$_4$[Fe(CN)$_6$]	blue solution turns reddish brown	Cu$_2$[Fe(CN)$_6$] forms 2Cu^{2+} + [Fe(CN)$_6$]$^{4-}$ === Cu$_2$[Fe(CN)$_6$]↓

[86] 制备

Table 1-3   Preparation[86] of KAl(SO$_4$)$_2$·12H$_2$O (on lab notebook)

m_{Al} / g	m_{NaOH} / g	$m_{K_2SO_4}$ / g	Yield/g	Outside observation of the product

Table 1-4   Preparation of KAl(SO$_4$)$_2$·12H$_2$O (on post-report)

m_{Al} / g	m_{NaOH} / g	$m_{K_2SO_4}$ / g	Yield/g	Outside observation of the product	Theoretical yield/g	Percent yield/%
1.00	2.04	3.46	15.0	small white crystal	17.6	85.2

[87] 标定

Table 1-5   Standardization[87] of a KMnO$_4$ solution (on lab notebook)

No.	1	2	3
m(Na$_2$C$_2$O$_4$)/g			
V_1(KMnO$_4$)/mL			
V_2(KMnO$_4$)/mL			
ΔV(KMnO$_4$)/mL			

Table 1-6   Standardization of a KMnO$_4$ solution (on post-report)

No.	1	2	3
m(Na$_2$C$_2$O$_4$)/g	0.2054	0.2153	0.2154
V_1(KMnO$_4$)/mL	0.00	0.04	0.08
V_2(KMnO$_4$)/mL	21.50	22.54	22.53
ΔV(KMnO$_4$)/mL	21.50	22.50	22.45
c(KMnO$_4$)/(mol·L^{-1})	0.02852	0.02856	0.02864
$\bar{c}$(KMnO$_4$)/(mol·L^{-1})		0.02857	
$\left| d_i \right|$ /(mol·L^{-1})	0.00005	0.00001	0.00007
$\overline{d_r}$ /%		0.2	

1.3.2　Guidelines for Pre-lab Report

It is critical that you complete a pre-lab report before coming to the laboratory. Here are some sections of the pre-lab report and guidelines to be followed.

(1) **Header:** Write the date, title of the experiment, your name, lab partner's name, and instructor's name.

(2) **Objectives:** Briefly and concisely describe the goals of the experiment and the methods involved[88].

[88] 概述涉及的实验目的和方法

(3) **Principles:** Explicitly review the theory behind the experiment, the related chemical reactions, and the methods used to investigate that theory[89].

[89] 简单描述涉及的原理、化学反应和研究方法

(4) **Answers to the pre-lab questions:** Interpret briefly.

(5) **Procedure:** Briefly and clearly describe each step with enough details that you or other people could read and conduct the experiment without additional guidelines[90]. A concise and clear procedure can prevent major errors during the experiment. Typically, create a flow chart indicating the major procedural steps and any potential pitfalls in the experiment[91]. A flow chart is particularly useful for planning time more effectively.

[90] 无需其他指南即可完成实验

[91] 画好流程图，标明实验中的主要步骤和潜在隐患

(6) **Key steps and notices**[92]**:** Use your own words to describe and summarize them briefly.

[92] 关键步骤和注意事项

1.3.3　Guidelines for Post-lab Report

The lab (post-lab) report should be completed independently, even if the experiment was done with partners. In the case of plagiarism, all students involved will receive a grade of zero[93]. **Note:** Do not give a copy of your lab report to other students.

[93] 若抄袭，所有涉及学生都判为零分

Here are some sections for the post-lab report and the guidelines to be followed.

(1) **Data record and analysis:** Numerical data should be presented in tables[94], containing only the facts.

[94] 数据记录与分析：数值型数据以表格呈现

(2) **Results and discussion:** Analyze the data and interpret whether the results can be accepted[95]. Discuss any errors and mistakes. Describe ways to improve the experiment and what you have learned from it.

[95] 结果与讨论：分析数据并解释(实验)结果是否可接受

(3) **Figures and graphs:** Label the axes on a graph, and ensure including all the units[96]. Number the figures consecutively, such as Figure 1, Figure 2, and so on.

[96] 图表：标注坐标轴的名称和单位

(4) **Questions:** All assigned questions should be briefly answered in this section.

(5) **References:** List references if needed.

扫一扫 视频1-3 实验记录本、预习报告和实验报告指南

1.4 Introduction to Titrimetric Analysis
滴定分析法概述

[97] 物质的分离、鉴定和定量测定
[98] 定性分析和定量分析

Analytical Chemistry plays an important role in our daily lives and involves the separation, identification, and quantification of matter[97]. Analytical Chemistry can be divided into three types of analysis: structural analysis, qualitative analysis, and quantitative analysis[98]. Quantitative analysis can be further divided into chemical analysis and instrumental analysis.

[99] 化学分析准确度高、操作简单、灵敏度低，用于常量分析
[100] 滴定分析和重量分析
[101] 滴定分析是通过测定标准溶液的体积得到试液浓度

Chemical analysis has the characteristics of high accuracy (E_r: 0.1%-0.2%), simplicity, and low sensitivity. It is suitable for macro analysis[99] ($m_s \geqslant 0.1$ g, $V \geqslant 10$ mL). Chemical analysis involves titrimetric analysis (also called titration analysis) and gravimetric analysis[100].

Titrimetric analysis involves the volume measurement of a standard solution and is used to determine the concentration of the analyte[101]. These techniques have remained in use due to their simplicity and high accuracy.

1.4.1 Important Terms

(1) Titrate(试液，待测液): the substance to be analyzed.

[102] 已知准确浓度的试剂

(2) Titrant(滴定剂): the reagent of accurately known concentration[102] which is added to the solution of the substance to be analyzed.

[103] 溶液的性质(如 pH 或电势)随滴定剂体积变化的曲线

(3) Titration curve(滴定曲线): a plot of solution property (such as pH or potential) versus titrant volume during a titration[103].

(4) Titration(滴定): the process of finding out the volume of the titrant required to react completely with an accurately known volume of solution.

(5) Indicator(指示剂): a species added to the solution to give an observable change at or near the endpoint.

(6) Endpoint (滴定终点): the point at which the reaction is finished or the color of indicator changes.

[104] 正好反应完全

(7) Stoichiometric point (化学计量点): the point at which the reaction is exactly complete[104].

[105] 指示剂产生的颜色变化通常比化学计量点稍微提前或推后

(8) Titration error (滴定误差): the smallest difference between stoichiometric point and endpoint. This difference exists because an indicator always produces the color change either a little before or after the stoichiometric point[105].

(9) Standard solution (标准溶液): (IUPAC) a solution of accurately known concentration.

1.4.2   Requirements for the Chemical Reactions

The titrate contains an unknown amount of chemicals, and reacts with the titrant in the presence of an indicator to signal the endpoint. Here are some requirements for the chemical reactions.

(1) The reactions take place quickly and essentially to completion.

(2) The reactions take place quantitatively, showing the stoichiometric relationship[106].

[106] 反应定量进行,具有化学计量关系

(3) There are no side reactions[107].

[107] 无副反应

(4) The endpoint should be well defined, either by an indicator or instrument[108].

[108] 可用指示剂或仪器判断滴定终点

1.4.3   Four Types of Titrimetric Analysis

Based on the types of chemical reactions involved, titrimetric analysis can be classified into four types: acid-base titration, redox titration, precipitation titration, and complexometric (complexation) titration[109].

[109] 按照涉及的化学反应类型,滴定分析分为酸碱滴定、氧化还原滴定、沉淀滴定和配位滴定

(1) Acid-base titration

$$NaOH + HCl = NaCl + H_2O$$

(2) Oxidation-reduction (Redox) titration

$$5C_2O_4^{2-} + 2MnO_4^- + 16H^+ = 10CO_2\uparrow + 2Mn^{2+} + 8H_2O$$

(3) Precipitation titration

$$Ag^+ + Cl^- = AgCl\downarrow$$

(4) Complexometric (complexation) titration

$$Zn^{2+} + H_4Y = ZnY^{2-} + 4H^+$$

1.4.4   Four Types of Titration Methods

Based on the types of performance, titrations can be classified into four types: direct titration, back titration, indirect titration, and replacement titration[110].

[110] 按照操作的类型,滴定方法分为直接滴定法、返滴定法、间接滴定法和置换滴定法

1. Direct Titration (also called simple titration)

In this process, the sample is titrated directly with a suitable standard solution or reagent. The amount of reagent consumed is quantitatively related to the amount of substance to be determined.

Note: A direct titration should meet all the requirements for the chemical reactions in 1.4.2.

2. Back Titration

Back titrations are usually used when the involved reaction is slow, no suitable indicator is available, or the endpoint is difficult to signal. In this process, two standards are used, causing two chemical reactions to take place.

The first standard is added in quantitative but excess to the sample, causing a chemical reaction between them. Then, the second standard is used to measure the remaining amount of the first standard[111].

3. Indirect Titration

In this process, the substance to be determined, containing a non-titratable form, is converted into a titratable compound through a chemical reaction[112]. For example, a $KMnO_4$ solution cannot titrate directly with a standard Ca^{2+} solution.

$$Ca^{2+} + C_2O_4^{2-} = CaC_2O_4\downarrow$$
$$CaC_2O_4 + 2H^+ = Ca^{2+} + H_2C_2O_4$$
$$2MnO_4^- + 5C_2O_4^{2-} + 16H^+ = 2Mn^{2+} + 10CO_2\uparrow + 8H_2O$$

therefore

$$5Ca^{2+} \rightarrow 5C_2O_4^{2-} \rightarrow 2MnO_4^-$$

4. Replacement Titration (substitution titration)

It is used when direct titration or back titration does not give a sharp endpoint. In this process, a good titratable component is released from the substance to be determined by the addition of suitable substances in excess[113], which can then be titrated directly.

$$Cr_2O_7^{2-} + 6I^- + 14H^+ = 2Cr^{3+} + 3I_2 + 7H_2O$$
$$I_2 + 2S_2O_3^{2-} = 2I^- + S_4O_6^{2-}$$

therefore

$$Cr_2O_7^{2-} \rightarrow 3I_2 \rightarrow 6S_2O_3^{2-}$$

1.4.5　Standard Solutions

1. Grade and Labelling of Chemicals

All chemical reagent bottles should have clear labels, including the reagent name in both Chinese and English, grade, formula, molecule weight, outside observation, physical properties, impurities[114] (Fig. 1-3).

Fig. 1-3　Labelling of chemical reagent

The grade indicates the purity of the chemical. There are many different grading standards, the most useful ones in general chemistry laboratories are shown in Table 1-7.

[111] (返滴定法)先加入定量过量的第一种标准物质，再用第二种标准物质去测定第一种剩余的量

[112] (间接滴定法)通过化学反应将不可滴定的形体转化成可滴定的化合物

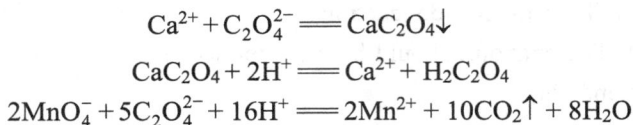

[113] (置换滴定法)加入过量的合适物质使原物质产生可滴定的组分

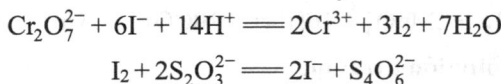

[114] (试剂瓶)标签涵盖试剂的中英文名称、等级、分子式、分子量、外观、物性、杂质

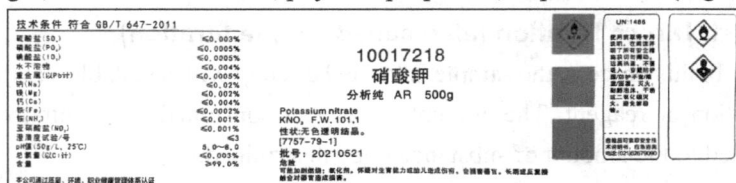

Table 1-7 Grade and labelling of chemicals

No.	Grade	Abbr.	Purity	Color of label	Applications
1	primary standard	PT	>99.9%	green	for preparing or standardizing standard solutions[115]
2	guaranteed reagents[116]	GR	>99.8%	green	for using in analytical chemistry
3	analytical reagents	AR	>99.7%	red	for laboratory and general use
4	chemical reagents	CP	>99.5%	blue	for general qualitative use and chemical preparation
5	lab grade	LB		brown	for general laboratory use

A good primary standard[117] meets the following main criteria: (1) high purity (above 99.9%), (2) high stability (i.e low reactivity), (3) high molecular weight (to reduce errors from mass measurements)[118], (4) low tendency to absorb moisture or CO_2 from the air.

Note: Higher-grade chemicals cost more than lower-grade chemicals, so choose the appropriate grade of reagents for each experiment.

2. Preparation of Standard Solution

According to IUPAC (International Union of Pure and Applied Chemistry), a standard solution is a solution of an accurately known concentration[119]. There are two common methods to prepare a standard solution: the direct method and the standardization method.

Ⅰ. The Direct Method

The direct method is suitable only for a primary standard. This solution is also called a primary standard solution[120]. The following four main steps are involved in this method.

(1) The mass of solute required is calculated and accurately weighed[121].

(2) The solute is dissolved in some water in a beaker.

(3) The solution is transferred into a volumetric flask[122].

(4) More water is added to reach the calibration mark.

Ⅱ. The Standardization Method (Indirect Method)

The concentration of the solution is determined by titration with a primary standard or another standard solution. This solution is also called a secondary-standard solution[123], such as standard HCl, NaOH, $KMnO_4$ and $Na_2S_2O_3$ solutions.

Note: The concentration of a secondary-standard solution will have greater uncertainty than that of a primary standard solution. If possible, standard solutions are best prepared by the direct method.

[115] 用于配制或标定标准溶液

[116] 保证级试剂

[117] 基准物质

[118] 分子量大(以减小称量误差)

[119] 标准溶液是指具有准确浓度的溶液

[120] 第一类标准溶液(直接用基准物质配制的溶液)

[121] 计算并准确称量所需的溶质质量

[122] 容量瓶

[123] 第二类标准溶液(用基准物质或标准溶液标定后得到浓度的溶液)

扫一扫　视频1-4　滴定分析法概述

1.5　Significant Figures and Data Treatment
有效数字和数据处理

When recording data in the laboratory and completing a laboratory report, one of the most confusing concepts for students is "significant figures". Generally speaking, significant figures are related to the precision of the measuring equipment[124]. For example, a length measured by a common ruler cannot be 16.6428721 cm.

[124] 有效数字反映了测量仪器的精度

1.5.1　Definition of Significant Figures

Significant figures are the digits in a measured quantity, including all digits that are known with complete certainty, plus the first digit with some uncertainty[125].

[125] 有效数字是测定数据中的数字，包括所有绝对确定的数字和第一位存疑数字

The uncertainty in the final digit is usually assumed to ±1 unless otherwise specified. For example, 8.2 mL implies a precision of only ±0.1 mL, but 8.20 mL implies a precision of ±0.01 mL, ten times the precision of the previous number. This indicates that 8.2 mL is measured by a graduated cylinder, while 8.20 mL is measured by a pipette[126].

[126] (要点)8.2 mL 说明精度只有±0.1 mL，可以用量筒量取；而 8.20 mL 需用移液管量取

[127] 称至 0.01 g

[128] (四位有效数字的精度)

Another example: a procedure may state, "weigh 15 g of NaCl to the nearest 0.01 g[127]". This statement indicates both the amount to be weighed (15 g) and the precision required for the experiment to have meaningful results (four-significant-figure precision)[128].

Precision and significant figures of some equipment are listed in Table 1-8.

Table 1-8　Precision (P) and significant figures (Sig. Fig.) of some equipment

Equipment	P	Sig. Fig.	Equipment	P	Sig. Fig.
graduated cylinder (10 mL)	0.1 mL	3	graduated pipette[129] (1-10 mL)	0.01 mL	3
volumetric pipette (1-10 mL)	0.01 mL	3	volumetric pipette[130] (10-50 mL)	0.01 mL	4
beaker	1 mL	2	volumetric flask[131]	0.01 mL	4
burette (10-50 mL)	0.01 mL	4	colorimetric cylinder[132]	0.01 mL	4
pH meter	0.01	2	conductivity meter[133]	0.001	3
top-loading balance	0.01 g	3	analytical balance[134]	0.0001 g	4
spectrophotometer	0.01	2	spectrophotometer[135]	0.001	3

[129] 分度吸量管

[130] 单标线吸量管

[131] 容量瓶

[132] 比色管

[133] 电导率仪

[134] 分析天平

[135] 分光光度计

1.5.2　Zero in the Significant Figures

It is often confusing to understand the meaning of zero in a number. Here are some rules.

(1) Non-zero digits and "middle" zeros are always significant[136]. "Middle" (imbedded in a number) zeros indicate that the reading on the instrument's scale is actually zero. For example, 124 has three significant figures, while 10204 has five significant figures.

(2) Leading zeros are never significant. Leading zeros only indicate the position of the decimal point relative to the first significant digit[137]. For example, 0.567 has three significant figures, and the first significant figure is in the tenth position after the decimal point. 0.00567 also has three significant figures, but the first significant figure is the thousandth position after the decimal[138].

(3) Trailing zeros in the decimal portion are significant[139]. For example, 0.0500 has three significant figures.

(4) Trailing zeros in a whole number are not significant. For example, 200 is considered to have only ONE significant figure, while 25000 has two. However, "200." with a decimal point has **THREE** significant figures[140].

(5) Scientific notation. To avoid confusion about whether zeros at the end of a number are significant, scientists use scientific notation. All numbers in scientific notation are significant[141]. For example, 5.00×10^{-2} has three significant figures.

1.5.3　Rounding Significant Figures

A number is rounded off to the required number of significant digits by dropping one or more digits from the right[142]. Here are some rules to round off the digits.

(1) When the digit after the rounding off digit is less than 5, drop all the digits on the right. For example, rounding 62.53475 to four significant figures should be 62.53.

(2) When the digit after the rounding off digit is greater than 5, add 1 to the rounding off digit and drop the remaining digits on the right. For example, rounding 62.53475 to five significant figures should be 62.535.

(3) When the digit after the rounding off digit is exactly 5, round the number so that the rounding digit becomes even[143]. For example, rounding 62.53475 to six significant figures should be 62.5348. Rounding 62.53485 to six significant figures should be 62.5348.

[136] 非零数字和中间的零都是有效数字

[137] 首位的零不是有效数字。它仅指示与第一位有效数字相关的小数点的位置

[138] 小数点后千分位

[139] 小数部分最后面的零都是有效数字

[140] 除非最后有小数点，否则整数最后的零都不是有效数字

[141] 科学计数法中的所有数字都是有效数字

[142] 修约就是从右边减去一位或多位数字以满足所需有效数字的位数

[143] 当被修约数字后面的数字正好是5,则将其修约成偶数

1.5.4　Mathematical Operations and Significant Figures

Correctly accounting for significant figures is important when performing mathematical operations[144] so that the resulting answers accurately represent the precision of all measuring equipment. In general, when combining measured numbers with different degrees of precision, the precision of the final answer cannot be greater than that of the least precisely measured number[145].

[144] 进行数学运算

[145] 将具有不同精度的测量值混合，最后结果的精度不能高于精度最低的测量值的精度

Here are some rules for performing mathematical operations.

(1) When performing **addition or subtraction**, the final answer can contain no more **decimal places** than the measured number with the largest absolute error[146]. For example, 78.**169** (three decimal places) + 62.**53** (**TWO** decimal places, having the largest absolute error) + 4.**0918** (four decimal places, having the smallest absolute error) = 78.**17** + 62.**53** + 4.**09** = 144.**79** (**TWO** decimal places).

[146] 进行加减运算时，最后结果的小数部分位数以绝对误差最大的数(小数点后位数最少的)来确定

(2) When performing **multiplication or division**, the final answer can contain no more **significant figures** than the measured number with the largest relative error[147]. For example, 12.536 (five significant figures, having the smallest relative error) × 2.184 (four significant figures) × 3.01 (**THREE** significant figures, having the largest relative error) = 12.5 × 2.18 × **3.01** = **82.0** (**THREE** significant figures).

[147] 进行乘除运算时，最后结果的有效数字位数以相对误差最大的数(有效位数最少的)来确定

[148] 用计算器或计算机运算时，(可以连续计算而不必每步修约)，但最后结果的有效位数需准确修约

(3) A calculator or computer can display a result with six, eight, or more digits, but this is not the final answer. You need to correctly round the final answer[148]. For example, 42.536 × **2.09** (three significant figures) = 88.90651= **88.9** (three significant figures).

1.5.5　Accuracy, Precision, and Measurement Error[149]

[149] 准确度、精密度和测量误差(说明：前面四段比较难理解，因此整段译成中文)

Accuracy refers to how close a measured number (value) is to the correct known value. Precision refers to how close the agreement is between repeated measured numbers, which are repeated under the same conditions. Measurements can be both accurate and precise, accurate but not precise, precise but not accurate, or neither (Fig. 1-4). High precision data indicates small **random error**[150] and leads experimenters to have confidence in their results. Highly accuracy data indicates minimal **systematic error**[151]. A well-designed experiment and a well-trained experimenter should yield data that is both precise and accurate.

[150] 随机误差

[151] 系统误差

准确度是指测量值与已知真实值的接近程度。精密度是指在相同条件下几次重复测量值之间的接近程度。测量结果可以准确度和精密度都高，准确度高但精密度不高，精密度高但准确度不高，或者两者都不高。精密度高的数据表明随机误差小，使实验者对结果有

信心。准确度很高的数据表明系统误差很小。一个精心设计的实验和一个训练有素的实验者能获得准确度和精密度都高的数据。

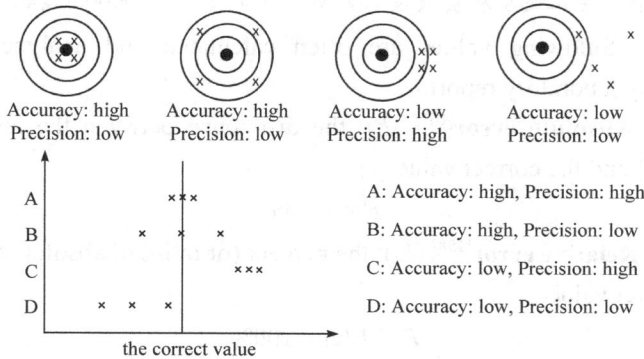

Accuracy: high　　Accuracy: high　　Accuracy: low　　Accuracy: low
Precision: low　　Precision: low　　Precision: high　　Precision: low

A: Accuracy: high, Precision: high
B: Accuracy: high, Precision: low
C: Accuracy: low, Precision: high
D: Accuracy: low, Precision: low

the correct value

Fig. 1-4　Accuracy and precision

Precision can be classified into repeatability and reproducibility[152].
Repeatability refers to the variation that occurs when all efforts are made
to keep conditions constant by using the same instrument and operator,
and repeating the measurements over a short time period. Reproducibility
refers to the variation arising by using the same measurement process
among different instruments and operators, and over longer time periods.

[152] 重复性和再现性

精密度可分为重复性和再现性。重复性是指仪器和操作者等条
件都相同，且在短时间内重复测量时产生的偏差。再现性是指仪器
和操作者都不同，但采取相同的测量过程，且在较长时间内完成而
产生的偏差。

Every measurement made in the laboratory is subject to error[153]. Two
types of errors will always occur: systematic errors and random errors.
Accuracy is related to systematic errors, usually expressed by error, while
precision is related to random errors, usually expressed by deviation[154].

[153] 误差

[154] 偏差

实验室完成的每个测量都注定有误差。通常出现两种类型的误
差：系统误差和随机误差。准确度与系统误差有关，通常用误差表
示。精密度与随机误差有关，通常用偏差表示。

Systematic errors are reproducible errors that can be corrected.
Examples include errors due to a miscalibrated[155] glassware or a balance
that consistently weighs lighter. Random errors are due to limitations in
measurement that are beyond the experimenter's control. Random errors
cannot be eliminated and lead to both positive and negative
fluctuations[156] in successive measurements. An example is the difference
in readings by different observers.

[155] 错误校准的

[156] 正负波动

系统误差是指重复的、可校正的误差。例如，由于错误校准的玻璃

器皿或一直偏轻的天平所造成的误差。随机误差是由测量的局限性造成的，而这些局限性是实验者无法控制的。随机误差无法消除，并在连续测量中产生正波动和负波动。例如，不同观察者得到的读数不同。

The following values are often calculated and analyzed in a chemistry laboratory report.

[157] 绝对误差：测量值 与真实值之间的差值

(1) **Absolute error**[157] (E): the difference between the measured value (x) and the correct value (x_T).

$$E = x - x_T$$

[158] 相对误差

(2) **Relative error**[158] (E_r): the percent (or ratio) of absolute error on the correct value.

$$E_r = E/x_T \times 100\%$$

Deviation gives information about how close or far a datum is from the other data of the set[159]. The higher the deviation, the further the data are from the mean.

[159] 偏差表明某个数据 与整组数据之间的远近 程度

[160] 绝对偏差

(1) **Absolute deviation**[160] is the difference between a single value and the average of the data.

[161] 相对平均偏差

(2) **Relative average deviation**[161] is defined as the mean deviation divided by the arithmetic mean, multiplied by 100%.

[162] 相对标准偏差

(3) **Relative standard deviation**[162] (RSD) is calculated as the ratio of standard deviation to the mean for a set of numbers.

$$\bar{x}(\text{average/mean}) = x_1 + x_2 + \cdots + x_n = \frac{1}{n}\sum_{i=1}^{n} x_i$$

$$d_i(\text{absolute deviation}) = x_i - \bar{x}$$

$$\bar{d}(\text{average deviation}) = \frac{1}{n}\sum |d_i| = \frac{1}{n}\sum |x_i - \bar{x}|$$

$$\bar{d_r}(\text{relative average deviation}) = \frac{\bar{d}}{\bar{x}} \times 100\%$$

[163] KMnO₄溶液的标定

Standardization of a KMnO₄ solution[163] by sodium oxalate ($Na_2C_2O_4$, a primary standard) is a common experiment in this manual. Now, take it for example to show how the data is recorded and analyzed (Table 1-9).

Table 1-9　Standardization of a KMnO₄ solution

No.	1	2	3
$m(Na_2C_2O_4)$ / g	0.2054	0.2153	0.2154
$V(KMnO_4)$ / mL	21.50	22.50	22.45
$c(KMnO_4)$ / (mol·L^{-1})	0.02852	0.02856	0.02864
$\bar{c}(KMnO_4)$ / (mol·L^{-1})		0.02857	
$|d_i|$ / (mol·L^{-1})	0.00005	0.00001	0.00007
$\bar{d_r}$ /%		0.2	

$$c_{KMnO_4} = \frac{2 \times m_{Na_2C_2O_4}}{5 \times M_{Na_2C_2O_4} \times V_{KMnO_4}} = \frac{2 \times 0.2054}{5 \times 134.0 \times 21.50} = 0.02852 \ (mol \cdot L^{-1})$$

$$\overline{d_r} = \frac{1}{3} \times \frac{\sum |d_i|}{\overline{c}} \times 100\% = \frac{1}{3} \times \frac{\sum |c_i - \overline{c}|}{\overline{c}} \times 100\%$$

$$= \frac{1}{3} \times \frac{0.00005 + 0.00001 + 0.00007}{0.02857} \times 100\% = 0.2\%$$

Note:

(1) Both value of $Na_2C_2O_4$ mass and $KMnO_4$ volume have four significant figures, so the concentration of $KMnO_4$ should have four significant figures[164], like 0.02857 mol·L^{-1}, not 0.0286 mol·L^{-1} or 0.029 mol·L^{-1}.

(2) The absolute deviations have only one significant figure, so relative average deviation should have only one significant figure[165], like 0.2%, not 0.15% or 0.152%.

(3) When **the leading significant figure is 9**, the value should have one more significant figure[166]. For example, **0.0965 has four significant figures**.

[164] 由于 $Na_2C_2O_4$ 质量和 $KMnO_4$ 体积均为 4 位有效数字，因此 $KMnO_4$ 浓度需保留 4 位有效数字

[165] ($KMnO_4$ 浓度)绝对偏差只有 1 位有效数字，因此相对平均偏差也只有 1 位

[166] 当首位有效数字为 9 时，该数值的有效数字位数多 1 位

扫一扫　视频 1-5　有效数字
视频 1-6　化学分析中的误差

（曾秀琼编写）

Chapter 2　Basic Operations and Common Instruments
基本操作和常用仪器

2.1　Cleaning and Drying Glassware
玻璃器皿的洗涤和干燥

2.1.1　Cleaning Glassware

Cleanliness of glassware in any laboratory is a key factor in obtaining correct results, making it extremely important to clean all glassware thoroughly.

1. General Notes

(1) NEVER assume any glassware is clean unless you have washed it yourself or have been informed.

(2) Wash and rinse glassware several times, using small amounts of detergent each time[1].

(3) Clean glassware immediately after each use to prevent residues from drying and becoming harder to remove. If a thorough cleaning is impossible at the time, soak the glassware in water[2].

(4) DO NOT wipe the inner walls of clean glassware with absorbent paper or cloth[3], as fibers or debris may be left behind.

2. Common Cleaning Methods

There are many types of residues (contaminants) that can be present on glassware, choose suitable methods to clean them.

(1) If glassware is new and intend for accurate measurements, such as a burette or a pipette[4], soak it for several hours in an acidic solution (1% HCl or HNO_3) before use, as new glassware tends to be slightly alkaline[5].

(2) If the contaminants are soluble or dust-like, they can be removed easily with a brush and water.

(3) If the contaminants are insoluble, such as grease[6], clean the glassware with detergent or cleaning powder (usually with an abrasive)[7]. Soak the glassware in a detergent or cleaning powder solution, scrub all parts of the glassware with a brush, and rinse thoroughly with tap water followed by deionized water[8]. **Note:** Heat the solution when needed.

[1] 洗涤原则是少量多次

[2] 若不能及时彻底洗涤，先将其浸泡在水中
[3] 不能用吸水纸或布擦干洗净器皿的内壁

[4] 滴定管或移液管

[5] 显弱碱性(含硅酸盐)

[6] 难溶性污物，如油脂
[7] 洗涤剂或去污粉(常含有摩擦剂)

[8] 先后用自来水和去离子水冲洗

(4) If the contaminants are acidic or alkaline, clean the glassware with an alkaline solution (such as NaOH, NaHCO₃) or an acidic solution (such as HCl). Rinse thoroughly with tap water followed by deionized water.

(5) If the contaminants are stubborn, such as coagulated organic matter[9], clean the glassware with a chromic acid cleaning solution[10]. First, remove excess water from the glassware, as water will reduce the cleaning effectiveness. Then, fill the glassware with the chromic acid cleaning solution and allow it to soak for some time (the duration depends on the amount of contamination on the glassware). After soaking, pour the cleaning solution back into its original container for reuse (chromic acid solution can be reused until it begins to turn a greenish color)[11]. Finally, rinse thoroughly with tap water followed by deionized water.

NOTE: Chromic acid cleaning solution should only be used when no other method can remove the contaminants[12]. It is highly acidic, corrosive and a carcinogen[13], so handle it with extreme care.

[9] 顽固污物，如凝结的有机物
[10] 铬酸洗液
[11] (铬酸洗液可反复使用，直至颜色变成绿色)
[12] 非必要不用铬酸洗液
[13] 致癌物

3. Testing the Cleanliness
Fill the glassware full of water, then invert it. A uniform film of water should run down the inner wall, with no droplets remaining on the walls[14].

[14] (洗净检测)均匀水膜沿内壁流下，不挂水珠

2.1.2 Drying Glassware
Here are some methods to dry glassware after cleaning.

(1) Air drying[15]: This is the most common and recommended method. Place the glassware upside down on a clean, non-absorbent surface, such as a drying rack[16] (Fig. 2-1), and allow it to air dry. Ensure the area is dust-free and away from potential contaminants.

[15] 自然晾干
[16] 滴(沥)水架

(2) Oven drying[17]: Place the glassware upside down in a 110ºC oven for several hours. Be sure to remove all plastic components from the glassware before drying.

[17] 烘箱烘干

(3) Airflow drying[18]: A glassware (airflow) dryer offers advantages such as fast drying, energy efficient, no water stains, easy to use, and simple maintenance (Fig. 2-2). Place the glassware upside down in the airflow dryer and set the temperature to around 110ºC.

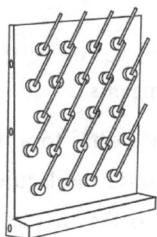

[18] 气流烘干器

Fig. 2-1 Drying rack   **Fig. 2-2 Airflow dryer**

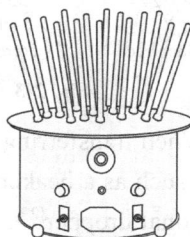

[19] 快速干燥：用乙醇润洗 2～3 次，再吹风使溶剂挥发

(4) Quick drying: Rinse the glassware with ethanol 2-3 times, then blow air into it to evaporate the solvent[19].

2.2 Transferring Reagents
试剂的取用

[20] 光敏性的

[21] 碱性溶液不能储存在带玻璃塞的试剂瓶中

[22] 细口瓶或滴瓶

Light-sensitive[20] chemicals should be stored in black or brown reagent bottles to avoid exposure to light. Alkaline solutions should not be stored in reagent bottles with glass stoppers[21]. Solid reagents are typically stored in wide-mouth bottles, while liquid reagents should be stored in narrow-mouth or dropping bottles[22].

[23] 瓶盖和瓶塞需倒放在实验台上

Note: Always check the label carefully before using any chemicals. Remove lids or stoppers and place them upside down on the laboratory bench[23]. Close the reagent bottle firmly after use.

2.2.1　Transferring Solid Reagents

[24] 药勺(角匙)

[25] 将卷好的纸送入试管中

When transferring a solid to a wide-mouth container, use a spoon[24]. When transferring a solid to a test tube, use a piece of glazed paper. Place the solid on the paper, roll it up with the solid inside, and insert the rolled paper into the test tube[25]. This method makes it easier to pour the chemicals into the test tube. **Note:** Avoid spilling out any chemicals and do not take excessive amounts.

2.2.2　Transferring Liquid Reagents

[26] 除非特别指明，让液体试剂瓶标签紧贴手心

[27] 保持滴管竖直

(1) Unless otherwise specified, hold the reagent bottle with your palm covering the label[26] to prevent any spilled liquid from damaging the label.

(2) When transferring liquid from a dropping bottle, always hold the dropper vertically[27] (Fig. 2-3). Never let the dropper touch any other surface.

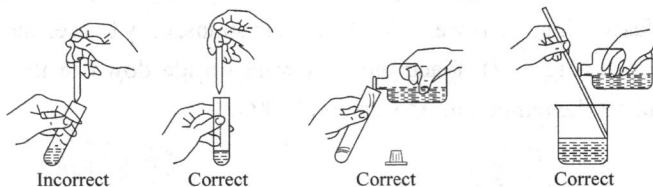

Incorrect　　Correct　　Correct　　Correct

Fig. 2-3　Transferring liquid reagents

[28] 用玻璃棒引流

[29] 不能用个人滴管

(3) When transferring liquid from a reagent bottle to a wide-mouth container, such as a beaker, use a glass rod to guide the liquid[28]. Never use a personal dropper[29]. Touch the glass rod to the inside wall of the beaker, then position the lip of bottle against the glass rod above the beaker. Pour the liquid along the glass rod into beaker (Fig. 2-3).

(4) When transferring liquid from a reagent bottle to a narrow-mouth container, such as a test tube, position the lip of the reagent bottle close to the mouth of the test tube[30].

[30] 将(细口)试剂瓶口紧贴试管口

2.3 Use of Volumetric Glassware
量器类玻璃器皿的使用

Introduction: In quantitative analysis, it is often necessary to make volume measurements with an error margin of approximately 0.2%[31]. This requires using glassware that can contain or deliver a volume known to about 0.01 mL. Therefore, report quantities greater than 10 mL to four significant figures[32].

[31] 定量分析中体积测量误差应小于 0.2%

[32] 大于 10 mL 的数值即为 4 位有效数字

Volumetric glassware is either calibrated "to contain" (In or TC) [33], such as volumetric flasks, or "to deliver" (Ex or TD)[34], such as graduated pipettes and burettes.

[33] 量入式(In 或 TC)
[34] 量出式(Ex 或 TD)

Volumetric glassware is calibrated with marks[35] used to determine the specific volume of liquid with varying degrees of accuracy. In a chemistry laboratory, four common types of volumetric glassware are the graduated cylinder, volumetric flask, burette, and pipette[36].

[35] 用刻度线标记

[36] 量筒、容量瓶、滴定管和移液管

Volumetric glassware can be classified as Class A, Class B or non-classified[37]. Class A glassware is more accurate and expensive.

[37] A 级、B 级或无等级

Note: Do not dry volumetric glassware using an oven or an airflow dryer. Instead, rinse with ethanol for quick drying.

Reading the volume: Due to surface tension and capillarity[38], liquids do not form a flat surface when in contact with the walls of a container. The curved surface is called a meniscus[39].

[38] 由于表面张力和毛细管作用

[39] 这个弯曲表面称为弯月面

The correct reading is taken at the lowest point of the meniscus with it at eye level[40]. It is often easier to see the meniscus by using a white card with a black strip on it (Fig. 2-4).

[40] 视线与弯月面最低点水平

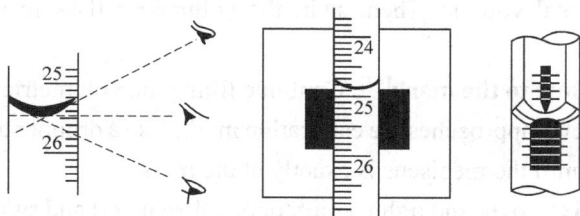

Fig. 2-4 Reading the volume

2.3.1   How to Use a Volumetric Flask

Introduction: Volumetric flasks (also called graduated flasks) are designed to contain a specific volume of liquid and are typically available in sizes ranging from 25 mL to 1 L. A volumetric flask is usually pear-shaped, with a flat base and a long neck. A ring-like mark on the neck indicates the calibration mark[41].

[41] 标线

1. Using a Volumetric Flask

To better understand how to use a volumetric flask, consider the example of preparing a solution from a solid reagent. Using a volumetric flask involves the following six steps.

[42] 检查漏液

[43] 盖上并旋紧瓶塞

(1) Check for leakage[42]: Add water to near the calibration mark of the volumetric flask, cover and tighten the stopper[43]. Keep the flask upside down for 2 minutes and check for any water around the stopper. If there is no leakage, set the flask upright, rotate the stopper about 180°, and invert the flask again to check (Fig. 2-5).

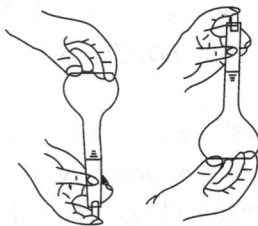

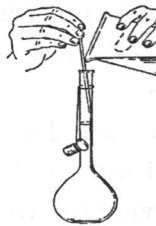

Fig. 2-5 Check   Fig. 2-6 Transfer   Fig. 2-7 Mix

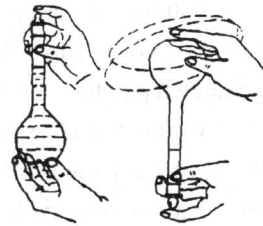

[44] 用分析天平称量所需质量的固体试剂

(2) Weigh and dissolve: Use an analytical balance to weigh the required mass of solid reagent[44], then transfer it to a beaker. Add some water and use a glass rod to stir the solution, ensuring the solid dissolves completely.

[45] 润洗烧杯内壁

(3) Transfer: Use a glass rod to transfer the solution to the volumetric flask (Fig. 2-6). Rinse the inside of the beaker[45] with a small amount of water 2-3 times, and transfer all the rinsed water into the volumetric flask.

[46] 预混

(4) Premix[46]: Fill the volumetric flask with water to approximately 2/3 of its total volume. Then, swirl the volumetric flask to premix the solution.

[47] 定容至标线

(5) Dilute to the mark[47]: Continue filling the volumetric flask until the liquid level approaches the calibration mark. Use a dropper to carefully add water until the meniscus is exactly at the mark.

[48] 翻转摇匀 10 次以上

[49] 高浓度的储备液

(6) Mix: Cover and tighten the stopper, then invert and swirl the flask at least 10 times to thoroughly mix the solution[48] (Fig. 2-7).

If using a concentrated stock solution[49] to prepare the solution, pipette the required amount of stock solution into the volumetric flask, then follow step (4) to step (6) as described above. This process is also called dilution[50].

[50] 稀释

2. Notes

(1) Check a volumetric flask for leakage before each use.

(2) Do not rinse a volumetric flask with the stock solution[51].

[51] 不能用储备液润洗容量瓶

(3) Do not transfer hot stock solution directly into a volumetric flask; allow the stock solution to cool down.

(4) Do not pour a solution directly into a volumetric flask. Always use a glass rod to transfer it.

(5) The upper side of the glass rod should not touch the mouth of the volumetric flask, as this could cause the solution to overflow[52].

[52] (从瓶口向外)溢出

(6) Do not add water above the calibration mark. If this happens, redo all the previous steps.

(7) Do not use a volumetric flask to store solutions for an extended period.

扫一扫　视频 2-1　容量瓶的基本操作

2.3.2　How to Use a Burette

Introduction: The burette, dating back to 1791, is one of the most accurate instruments for measuring and delivering small amounts of liquids. A burette consists of a long graduated glass tube with a stopcock near the tip to control the flow of liquid[53]. The minimum graduation on a 25 mL or 50 mL burette is 0.1 mL, so the volume should be read to the closest 0.01 mL. Modern burettes are made from borosilicate glass with a Teflon stopcock[54] (Fig. 2-8). One end has a handle for opening and closing, while the other has a tightening nut[55], a rubber o-ring, and a Teflon spacer (ring)[56]. A burette with a Teflon stopcock can be used for most liquids, including acidic or alkaline solutions.

[53] 滴定管由一根带有刻度的长玻璃管和底部带有可控制流速的旋塞组成

[54] 由硅硼玻璃制成，带有特氟龙(聚四氟乙烯)旋塞

[55] 旋紧螺母

[56] 垫片

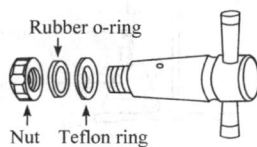

Rubber o-ring

Nut　Teflon ring

Fig. 2-8　Teflon stopcock

1. Using a Burette

Using a burette involves the following seven steps.

(1) **Check for leakage**[57]: Fill the burette with water to near the "0" mark. Rotate the stopcock to fill the tip of the burette with water. Dry the outside and tip of the burette with absorbent paper. Keep the burette upright for about 2 minutes, observe whether there is water around the stopcock or tip[58].

[57] 检查漏液

[58] 活塞或尖嘴附近

[59] 用少量待装液(滴定剂)润洗滴定管 2～3 次，每次从旋塞放出溶液

[60] 避免滴定剂被稀释

[61] 用手指轻弹或快速转动旋塞

[62] 若滴定管尖嘴残留液滴，用吸水纸除去

[63] 滴定架

[64] 用左手(反向)转动旋塞慢慢加入液体，同时用右手旋摇锥形瓶

[65] 临近终点，每次加半滴。用锥形瓶靠下悬挂的半滴

[66] 对深色溶液，读取液面的最高点

(2) Clean and rinse: First, clean the inside of the burette with tap water or detergent. Then rinse the burette 2-3 times with a small amount of the solution to be filled (the titrant), draining the titrant through the stopcock after each rinse[59]. Rinsing with the solution is to prevent dilution of the titrant[60].

(3) Fill the burette with the titrant and remove any bubbles: Fill the burette with the titrant above the "0" mark. Tap the burette with your fingers or quickly rotate the stopcock[61] to remove any air bubbles inside.

(4) Read the initial volume: Drain the solution until the meniscus is at or slightly below the "0" mark. If there is a drop of solution clinging to the burette tip, remove it with absorbent paper[62]. Read and record the initial volume of the burette at the meniscus.

(5) Titrate:

1) Carefully clamp the burette to a burette stand[63]. Insert the burette tip about 1 cm deep into the Erlenmeyer flask (Fig. 2-9).

2) Rotate the stopcock with your left hand to slowly add the titrant, while swirling the Erlenmeyer flask with your right hand[64] (Fig. 2-10). At the beginning, you can titrate at a slightly faster speed.

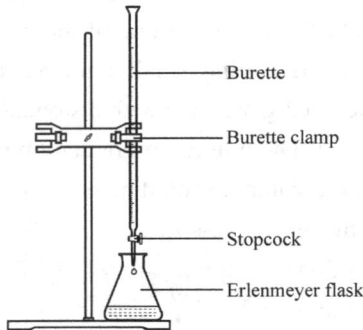

Fig. 2-9　Burette and stand　　　　Fig. 2-10　Titration operations

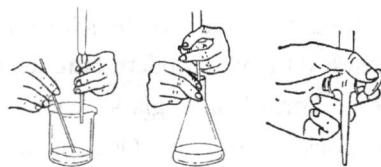

3) When approaching the endpoint, add the titrant drop by drop, swirling the Erlenmeyer flask after each drop.

4) Near the endpoint, add half a drop of the titrant each time. Touch the Erlenmeyer flask to the hanging half drop[65]. Rinse the inner walls with deionized water and swirl to mix well.

(6) Read the final volume: Hold the top of the burette vertically with your fingers. Read and record the final volume of the burette. **Note:** For colorless or light-colored titrant, ensure your eyes are at the level of the meniscus. For dark-colored titrant, ensure your eyes are at the level of the highest point of the solution[66].

(7) Clean up: Empty the remaining titrant into the designated container. Rinse the burette several times with tap water followed by deionized water. Carefully clamp the burette upside down on the burette stand.

2. Notes

(1) Ensure that the stopcock is tight enough to prevent leakage, but loose enough to rotate freely[67].

[67] 确保旋塞既不漏液又能自由旋转

(2) Close the stopcock whenever the burette is prepared.

(3) Sit on a chair or stand up straight to perform the titration, never squat down, hunch over, stand in horse stance, or kneel[68].

[68] 不能蹲着、弓着背、扎马步或跪着(滴定)

(4) Always transfer the titrant directly to the burette; do not pour it into any other glassware, such as a beaker. Pouring it into other glassware may change the concentration and purity of the titrant.

(5) Do not titrate too quickly, the titrant should not flow down in a continues stream[69].

[69] 滴定液(应成串而)不能成(线状)水流流下

(6) Whenever reading the volume, remove the burette from the clamp and hold the top of the burette vertically with your fingers[70].

[70] 手指竖直捏住管口

(7) To perform additional titrations, refill the titrant to around "0" mark each time.

扫一扫   视频 2-2   滴定管的基本操作

2.3.3   How to Use a Pipette

Introduction: Pipettes are used in laboratories to measure and transport precise volumes of liquid, typically ranging from less than 1 mL to about 100 mL. There are two common types of pipettes (Fig. 2-11). One is the Mohr or **graduated pipette**, which has scale marks and is used to deliver varying volumes of liquid[71]. Another is the **volumetric pipette**, which has one graduation mark and is used to deliver only a single specific volume of liquid accurately. A volumetric pipette is commonly called a bulb pipette[72].

[71] 分度吸量管(刻度移液管),沿着管身标有刻度,用于移取不同体积的液体
[72] 单标线吸量管只能准确移取一个体积的液体,俗称胖肚移液管

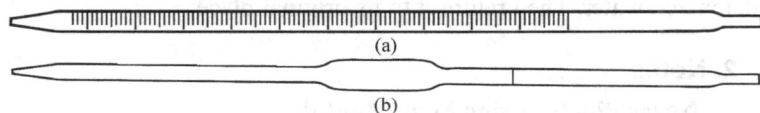

Fig. 2-11   Graduated pipette (a) and volumetric pipette (b)

1. Using a Pipette

Using a pipette includes the following five steps. To better understand, consider the example of transferring the solution from a reagent bottle to an Erlenmeyer flask.

(1) Check: Check for the intactness of the tip and top of the pipette. Check the marked capacity and scale[73]. Also, check if it is marked as "blow out"[74].

(2) Clean and rinse: Wash the pipette with deionized water 2-3 times. To avoid diluting the solution to be transferred[75], rinse the pipette with the solution 2-3 times. Each time, draw a small amount of the solution into the pipette, hold it horizontally, and roll it so that the solution coats the inside surface[76]. Finally, empty (drain) the liquid through the tip.

(3) Draw up the solution[77]: Hold the top of pipette with one hand and immerse the tip into the solution, ensuring the tip remains below the surface at all times. With the other hand, squeeze the rubber suction bulb before pressing it onto the top of pipette[78], then slowly release the rubber suction bulb to draw up the solution until the meniscus is several centimeters above the calibration mark. Quickly remove the rubber suction bulb and press the top of the pipette with your index finger[79]. Raise the pipette, drawing the tip to the mouth of the reagent bottle. Gently twist your index finger to allow the solution to flow down[80] until the bottom of the meniscus is exactly at the mark. Press the top of pipette firmly again to prevent further flow.

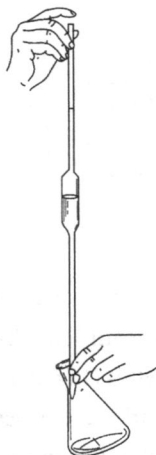

(4) Release the solution: Remove the pipette from the mouth of the reagent bottle, transfer it to the Erlenmeyer flask. Touch the tip of pipette to the inner wall of the Erlenmeyer flask, then remove your index finger to release the liquid[81] (Fig. 2-12). Wait for 10 seconds after releasing all the liquid, then rotate the pipette one circle.

Fig. 2-12　Pipette

(5) Clean up: Wash the pipette several times with tap water followed by deionized water. Then return it to its original place.

2. Notes

(1) **Never Pipette Using Your Mouth!**

(2) **Never** use your thumb to press the top of pipette.

(3) When drawing up the solution, ensure the tip of pipette remains below the surface at all times to prevent air from being drawn into the pipette[82].

[73] 检查标记总容量和分刻度

[74] 检查是否标有 "吹" 字

[75] 为避免待转移溶液被稀释

[76] 平端移液管并转动，使溶液遍及其内表面

[77] 吸液

[78] 先挤压洗耳球，再将其伸入移液管管口内

[79] 快速放下洗耳球，用食指堵住管口

[80] 轻轻转动食指使溶液流下

[81] 将移液管尖嘴抵住锥形瓶内壁，松开食指，使溶液流出

[82] 确保尖嘴全程在液面以下，以防吸空

(4) After releasing the solution, do not blow out the remaining solution in the tip unless the pipette is marked as "blow out"[83], as it has been calibrated to retain that amount.

(5) For graduated pipettes, the basic operations are the same. However, always start from the "0" mark and avoid using the mark near the tip[84].

[83] 除非标有"吹"字,否则放液后不能吹出尖嘴处的残留液滴

[84] 靠近尖嘴的刻度

扫一扫　视频 2-3　移液管的基本操作

2.3.4　How to Use an Automatic Pipette (Micropipette)

Introduction: Automatic pipettes, also known as micropipettes, are laboratory equipment used to accurately and quantitatively transfer volumes of liquid in the microliter range[85]. In 1957, Heinrich Schnitger invented the first micropipette, which was a fragile glass capillary tube requiring expert handling. In 1960, Dr. Warren Glison introduced a piston-based adjustable micropipette to solve this problem.

[85] (移液器)用于准确定量转移毫升级的液体

Micropipettes are available in single-channel (Fig. 2-13) and multi-channel versions, adjustable (variable) volume and quantitative (fixed) volume, mechanical and electronic, air displacement and positive displacement types[86]. Common micropipette ranges include: P10 (1.0 to 10 μL), P20 (2.0 to 20 μL), P200 (20 to 200 μL), P1000 (100 to 1000 μL).

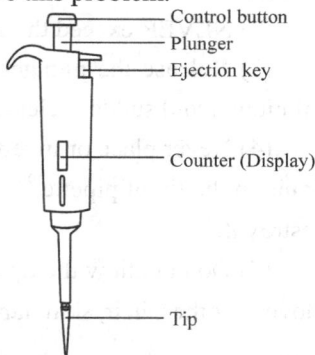

Control button
Plunger
Ejection key

Counter (Display)

Tip

Fig. 2-13　Micropipette

[86] 单道和多道、可调(变化)体积和定量(固定)体积、手动和电动、空气置换和外置活塞等种类

1. Using an Automatic Pipette

Using a micropipette includes the following five main steps.

(1) Set the desired volume: When adjusting the volume from high to low, turn directly to the desired volume. When adjusting from low to high, first surpass the desired volume by at least 1/3 turn, then reverse to the desired volume[87]. **Note:** Operate the volume adjustment knob with light force and never turn the knob if you feel resistance[88], as turning towards the direction of resistance can damage the pipette.

(2) Put on a tip[89]: Insert the pipette vertically into the tip, then slightly turn it from left to right to ensure it locks firmly.

[87] 先超出所需体积至少1/3 圈,再回调

[88] 轻轻转动体积调节旋钮,若感觉到阻力则不再转动

[89] 安装枪头

[90] 摁压活塞至第一停点，将枪头伸入液面下

[91] 不能将枪体浸入液体中

[92] 将枪头抵住容器内壁成 45°

[93] 摁至第二停点清空尖嘴处残留液滴

[94] 弹射键

[95] 将体积调至最大

(3) Suck up (aspirate) the liquid: The plunger will stop at two different positions when pressed. Press the plunger to the first stop, and immerse the tip in the liquid[90]. Keep the pipette in a vertical position, and slowly release the plunger. After sucking up the liquid, keep the tip in the liquid for 1-3 seconds, then remove it. **Note:** Never submerge the pipette shaft in the liquid[91].

(4) Release the liquid: Place the tip against the inner wall of the receiving container at a 45°[92]. Gently press the plunger to the first stop to release the liquid. Wait a few seconds, then press the plunger to the second stop to discharge any remaining liquid in the tip[93].

(5) Clean up: Press the ejector button[94] to discard the tip into a designated waste container. Adjust the volume to its fully extended position[95]. Place the pipette vertically on the pipette rack.

2. Notes

(1) Always select the SMALLEST pipette that will handle the volume you wish to transfer to achieve the greatest accuracy. Accuracy decreases when using unnecessarily large pipettes for small volumes[96].

[96] 移液器量程越大，准确度越小

[97] 不要超出移液器最高或最低量程

(2) NEVER exceed the upper or lower limits of the pipettes[97].

(3) Release the plunger slowly and steadily when drawing up the solution; avoid sudden release.

(4) Never place or invert the pipette shaft horizontally when there is liquid in the tip of pipette[98], as liquid may run down into the pipette and destroy it.

[98] 当枪头内有液体时，不能将移液器平放或倒置

(5) Do not allow the tip of the pipette to touch any object (including gloves, clothes, hair, skin, lab bench).

扫一扫　视频 2-4　移液器的基本操作

2.4　Separation Techniques
分 离 技 术

For an inorganic and analytical chemistry laboratory, the common techniques for separating solid-liquid mixtures are decantation, filtration, and centrifugation[99].

[99] 倾析、过滤和离心

2.4.1　Filter Paper

Filter paper may be made from natural cellulose, nylon, glass or quartz fibers[100]. They can be categorized as either qualitative or quantitative.

　　Qualitative filter paper is used for general purpose filtration applications where subsequent analysis is not critical. It generally contains more residual ash[101].

　　Quantitative filter paper requires high precision with low impurities and ash, typically less than ≤ 0.009% of ash[102]. It is used for quantitative analysis.

　　Table 2-1 shows the requirements of quantitative filter paper. See GB/T 1914—2017 "Chemical Analysis Filter Paper"[103] for more information.

[100] 天然纤维素、尼龙、玻璃或石英纤维

[101] (定性滤纸)通常含有更多残留灰分

[102] (定量滤纸灼烧后)灰分含量< 0.009%

[103] GB/T 1914—2017《化学分析滤纸》

Table 2-1　Requirements of quantitative filter paper

项目		要求								
		优等品			一等品			合格品		
		201 型	202 型	203 型	201 型	202 型	203 型	201 型	202 型	203 型
定量	g/m²	80.0 ± 4.0			80.0 ± 4.0			80.0 ± 5.0		
分离性能	—	合格								
滤水时间	s	≤35	35～70	70～140	≤35	35～70	70～140	≤35	35～70	70～140
干耐破度 >	kPa	85	90	90	85	90	90	80	85	85
湿耐破度 >	mm 水柱	130	150	200	120	140	180	120	140	180
抗碱性 >	%	95.0	95.0	95.0	95.0	95.0	95.0	93.0	93.0	93.0
灰分 <	%	0.009			0.010			0.011		
水抽提液 pH	—	5.0～8.0								
交货水分	%	7.0 ± 3.0								

2.4.2　Decantation

　　If a solid settles well, the liquid can sometimes be poured off (decanted). Decantation is the process of removing the liquid layer from the top of the solid or liquid layer below. This can be done by tilting the mixture and carefully pouring off the top layer[104]. However, this process is often not efficient, as a thin layer of the remaining liquid is difficult to remove completely.

[104] (倾析法)将混合物倾斜，小心倒出上层溶液

2.4.3　Gravity Filtration

　　Gravity filtration is a method of separating solids from liquids using the force of gravity to pull liquid through filter paper[105]. Its advantages include easy setup, minimal equipment required, and simple operations.

[105] (常压过滤)利用重力作用使液体通过滤纸

However, its disadvantage is that it is not suitable for filtering large volumes or fine particles.

Note: Gravity filtration is relatively slow compared to other filtration methods. When needed, prepare a water column in the funnel neck to help accelerate gravity filtration through the attraction action of the water column[106].

[106] 在漏斗颈部做一段水柱，利用水柱抽吸作用加快常压过滤

1. How to Perform a Gravity Filtration

Gravity filtration includes the following four main steps (Fig. 2-14).

(1) Fold: Fold a piece of filter paper in half twice to form a cone when unfolded[107].

[107] 对折滤纸两次，打开即为一个圆锥体

(2) Tear off: Tear off a corner of the filter paper to create a tighter seal between the filter paper and the funnel wall[108]. **Note:** The corner should be torn from the side with three-layers.

[108] 使(三层处)滤纸和漏斗内壁之间更加密封

(3) Set and wet: Place the filter paper in the funnel, fitting it as closely as possible. Wet the filter paper with the solvent (usually deionized water in this manual) and press it against the funnel around its circumference to seal the space between the filter paper and the funnel wall[109].

[109] 用溶剂润湿滤纸，按压滤纸与漏斗壁紧贴

(4) Transfer: Pour the solid-liquid mixture onto the filter paper, allowing the filtrate to collect in a container[110] (like a beaker). Use a glass rod to guide the flow of liquid and prevent splashing.

[110] 使滤液流入容器中

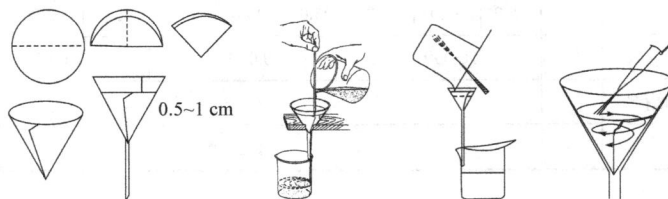

0.5~1 cm

Fig. 2-14　Gravity filtration operations

2. Gravity Filtration in Gravimetry

In a gravimetry (gravimetric analysis) experiment, the following precautions should be taken.

[111] (重量分析法)先将大部分上清液倾析至滤纸上

(1) The first step is to decant the majority of the supernatant into the filter paper[111] without transferring the precipitate. This helps prevent the filter paper from clogging[112] at the beginning of the filtration process.

[112] 以防滤纸(孔)被堵

(2) Rinse the precipitate in the beaker several times. Each time, decant the rinsed solution into the filter paper[113].

[113] 将洗涤后的溶液倾析至滤纸上

(3) Use a stream of rinsing solution to transfer the precipitate to the filter paper. Any precipitate that clings to the walls of the beaker can be transferred using the small corner of filter paper that was previously torn off[114].

[114] 黏在烧杯壁上的沉淀用前面撕下的滤纸角转移

2.4.4   Suction/Vacuum Filtration

A suction (vacuum) pump is generally used to create a pressure difference across the filter paper, effectively drawing fluid through it[115]. This technique significantly speeds up the filtration and can also aid in drying the solid.

[115] (减压过滤/抽滤)真空泵在滤纸两侧产生压力差使液体通过

1. Main Instruments

The main instruments used in suction filtration are a vacuum pump (or water-circulation pump), a Buchner funnel, and a suction flask (filter flask)[116] (Fig. 2-15).

[116] 真空泵(或循环水真空泵)、布氏漏斗、抽滤瓶

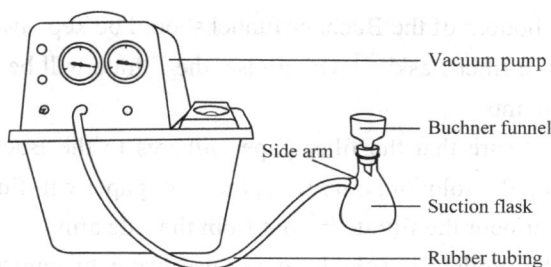

Fig. 2-15   Suction filtration apparatus

2. How to Perform a Suction Filtration

Suction filtration includes the following six main steps.

(1) Set up the apparatuses: Secure the filter flask with an iron ring to prevent from tipping over or moving[117]. Attach rubber tubing to the side arm of the flask[118] and connect the tubing to a vacuum pump. Ensure the rubber tubing is not bent, as this can lead to poor suction. The bottom of the Buchner funnel should be kept away from the side arm.

[117] 用铁圈固定抽滤瓶，使其不会倒翻或移动
[118] 将导管和支管连接

(2) Prepare the filter paper: Prepare a piece of suitable filter paper that is slightly smaller than the inside diameter of the Buchner funnel, ensuring it covers all the porcelain holes without extending over the edge of the funnel[119].

[119] 既能盖住所有瓷孔，又不能超出其边缘

(3) Wet the filter paper: Wet the filter paper with solvent (usually deionized water in this manual). Turn on the vacuum pump to ensure the filter paper adheres tightly to the Buchner funnel[120].

[120] 确保滤纸紧贴布氏漏斗

(4) Filter the mixture: While pumping, pour the solid-liquid mixture into the Buchner funnel in portions. Use the solution in the filter flask ("mother liquor") to help transfer the remaining solid residue into the funnel[121].

[121] 用抽滤瓶内的溶液(母液)转移残留固体

(5) Rinse the solid product[122]: After filtration, impurities may remain on the surface of the solid product. Solvents, such as ethanol, can be used to rinse the product.

[122] 洗涤固体产品

1) Remove the rubber tubing.

2) Add rinsing liquid until it just covers all portions of the solid. Carefully break up any solid clumps with a spoon or a glass rod, taking care not to tear the filter paper[123].

[123] 用药匙或玻璃棒压散块状固体，注意不要划破滤纸

3) Wait for a while. Connect the rubber tubing. Suction for a while to remove the solvent and dry the product.

(6) Get the solid product: Remove the rubber tubing and turn off the vacuum pump. Use a spatula or a glass rod to transfer the solid product to a pre-weighed watch glass[124].

[124] 已称量的表面皿

3. Notes

(1) The bottom of the Buchner funnel should be kept away from the side arm of the filter flask[125]. Otherwise, the filtrate will be sucked into the vacuum pump.

[125] 布氏漏斗最下端远离抽滤瓶支管

(2) Make sure that the filter paper adheres to the Buchner funnel before pouring the solution; otherwise, the filter paper will float up[126].

[126] 漂浮起来

[127] 滤液

(3) Never pour the filtrate[127] out from the side arm.

(4) If the product is soluble, never use water to transfer the solid residue into the Buchner funnel or to rinse the product!!! Otherwise, the product will dissolve and "disappear"!!!

(5) When the filtration is finished, remove the rubber tubing from the filter flask before turning off the vacuum pump. Otherwise, the lower pressure in the filter flask may cause water to be sucked back into the mother liquor[128].

[128] 否则，抽滤瓶内的低压会使水倒吸至母液中

扫一扫　视频 2-5　减压过滤的基本操作

2.5　Common Instruments
常 用 仪 器

2.5.1　Analytical Balance

1. Introduction

[129] 电子天平
[130] 机械天平
[131] 分析天平

Modern electronic balances[129] offer convenience in weighing and result in fewer errors or failures compared to mechanical balances[130] (Fig. 2-16). An analytical balance[131] is designed to measure mass in the milligram (mg) and even sub-milligram range (0.1 mg).

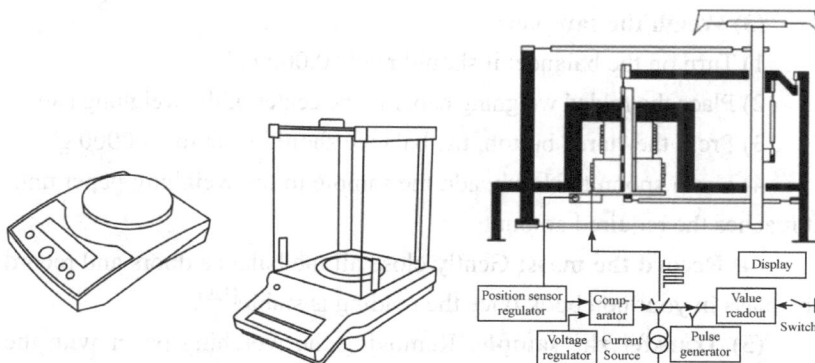

Fig. 2-16　Weighing instruments

An electronic analytical balance measures the force needed to counter the mass being measured rather than the actual masses. The balance compensates for gravitational differences by using an electromagnet to generate a force that counters the sample being measured[132].

Briefly, the balance measures the downward force acting on the weighing pan, which is then converted into an electrical signal and displayed on a digital screen. The weighing pan is enclosed to prevent air currents from affecting the measurement[133]. The "tare" function[134] makes it convenient to weigh and record mass accurately.

2. Weighing Methods

Ⅰ. Direct Weighing

Direct weighing is placing the substance directly on the weighing pan and measuring its mass, such as weighing a crucible in gravimetric analysis[135].

Ⅱ. Weighing by Addition

Weighing by addition is adding the substance until the required amount is reached (Fig. 2-17). It is suitable for weighing substances that are not hygroscopic and stable in air[136], such as metals and minerals. The process includes the following six steps.

(1) **Fold a weighing paper:** Take a piece of weighing paper, and carefully fold it into a dustpan shape to prevent solids from spilling out[137].

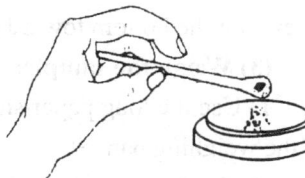

Fig. 2-17　Direct weighing

(2) **Check the leveling and cleanliness of the balance:** If the leveling bubble is not centered, adjust it by turning the leveling screws on the bottom of the balance toward the back[138].

[132] 电子分析天平测量的是抵消被测质量所需的力，而不是实际质量。天平利用电磁体产生与被测样品相反的力来补偿重力差

[133] 秤盘被封闭以避免气流对称量的影响

[134] "去皮"功能

[135] 如重量分析法中称量一个坩埚的质量

[136] (增量法称量)适合称量不吸湿且在空气中稳定的物质

[137] 将称量纸折成簸箕状，防止固体试剂撒出

[138] 若水平气泡不居中，调节天平底部后面螺丝

(3) Weigh the sample:

1) Turn on the balance; it should read "0.0000 g".

2) Place the folded weighing paper in the center of the weighing pan.

3) Press the "tare" button, the balance should reset to "0.0000 g".

4) Use a spoon to slowly add the sample to the weighing paper until it reaches the required amount.

(4) Record the mass: Gently close all the balance doors and record the mass in your notebook once the reading is stable[139].

[139] 小心关上所有天平门,待读数稳定后记录质量

(5) Transfer the sample: Removing the weighing paper with the sample from the balance and pour the sample completely into a clean receiving container.

(6) Clean up: Turn off the balance, clean the weighing pan, and close all the balance doors.

Ⅲ. Weighing by Difference or Subtraction

Weighing by difference, also known as the subtraction method[140], excels in accuracy and contamination control, making it ideal for the utmost accuracy. In this method, the sample weight is determined by the difference between the two weights of the weighing bottle before and after transferring the sample[141].

[140] 差减法称量，又称减量法称量

This method is often used for samples that are hygroscopic, easily oxidized or can react with the substance in the air[142]. These samples should be stored in a covered weighing bottle for protection.

[141] 称出的样品质量是倒出样品前后两次称量瓶质量之差

[142] 易吸湿，易被氧化或与空气中的物质反应

(1) Take a weighing bottle: Using a long paper strip to remove the weighing bottle with the sample from the desiccator[143] (Fig. 2-18). Place the weighing bottle in a clean dry petri dish[144] to prevent contamination.

[143] 用一条长纸带从干燥器里取出称量瓶

[144] 培养皿

Fig. 2-18   Weighing bottle

(2) Check the leveling and cleanliness of the balance: If the leveling bubble is not centered, adjust it by turning the leveling screws on the bottom toward the back.

(3) Weigh the sample:

1) Use the long paper strip to place the weighing bottle in the center of the weighing pan[145].

[145] 秤盘

2) Gently close all the balance doors, tare the balance, and ensure it reads "0.0000 g".

3) Use the long paper strip to remove the weighing bottle from the balance, and position it above a receiving container. Using a short paper strip, hold and remove the stopper. Tap the stopper against the upper side of the weighing bottle, the samples should fall down into the receiving container[146].

[146] 用瓶盖敲击称量瓶上端,样品落入承接容器内

4) Slowly upright the weighing bottle and cover it with the stopper.

5) Place the weighing bottle back on the weighing pan and read the mass. The mass should be a negative number due to taring[147]. **Note:** Do not close the balance doors at this time.

6) Repeat steps 3), 4) and 5) until the required amount.

(4) Record the mass: Gently close all the balance doors and record the mass once the reading is stable.

(5) Clean up: Return the weighing bottle to the desiccator, turn off the balance, clean the weighing pan, and close all the balance doors.

3. Notes

(1) Specific finger sleeves[148] can be used to hold the weighing bottle.

(2) Analytical balances are precise and expensive instruments; only weighing paper or weighing bottles should be placed on the weighing pan. Never place wet glassware, such as beaker or Erlenmeyer flask, on the weighing pan.

(3) Always close all doors when reading the mass or taring the balance[149], as air movement can impact the accuracy of the mass measurements.

(4) Close the doors gently because any vibration will damage the balance. Clean up immediately if samples spill.

(5) Never touch the weighing bottle with bare hands, as moisture or dust from your hands could affect the measurement.

[147] 质量应为负数,因为去皮了

[148] 特定的手指套

[149] 读数或去皮时,需关闭所有天平门

扫一扫   视频 2-6   分析天平的基本操作

2.5.2   Spectrophotometer

1. Introduction

A spectrophotometer is an instrument that measures the amount of light absorbed by a sample (Fig. 2-19). It is used in various aspects of chemical, biochemical research, and industrial technology. A UV-visible spectrophotometer[150] uses light in the ultraviolet range (185-400 nm) and

[150] 紫外-可见分光光度计

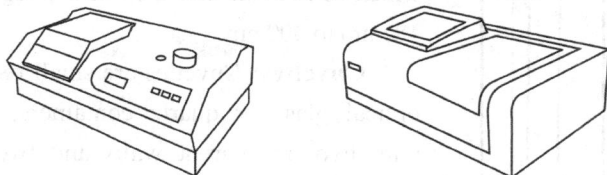

Fig. 2-19   Spectrophotometer

visible range (400-760 nm) of the electromagnetic spectrum.

2. Measuring Principles

When a light beam passes through a solution, it selectively absorbs a certain wavelength of light. The color presented by the solution is the complementary color of its strong absorption color. The Lambert-Beer Law states that there is a linear relationship between the absorbance and the concentration of a sample[151].

$$A = -\lg T = -\lg(I_t/I_0) = \varepsilon bc$$

where T is transmittance[152], A is absorbance of light, I_0 is intensity of incident light[153], I_t is intensity of transmitted light, ε is molar absorptivity (or molar extinction coefficient)[154] (L·mol^{-1}·cm^{-1}), c is concentration of the sample solution (mol·L^{-1}), b is light path length (cm), usually the length of a cuvette[155].

3. Instrument Components

Most spectrophotometers are composed of five components (Fig. 2-20). (1) a stable light source, (2) a monochromator[156], which isolates a specific region of the spectrum for measurement, (3) a sample container[157], (4) a detector, which converts radiant energy into a measurable electrical signal[158], and (5) a signal processing and readout unit (digital display)[159], usually consisting of electronic hardware and a computer.

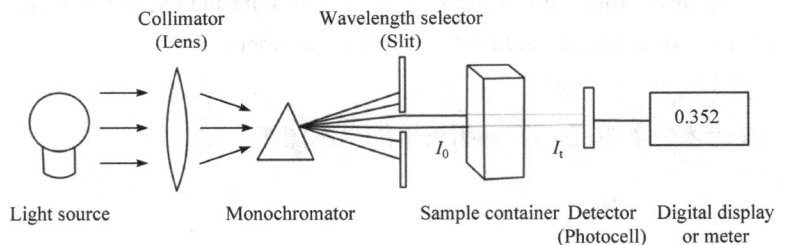

Fig. 2-20　Components of spectrophotometer

Light source: Visible spectrophotometers use a tungsten halogen lamp[160] as the light source, which emits light from about 350 nm in the UV region up to over 2500 nm in the near-infrared region. Ultraviolet (UV) spectrophotometers use hydrogen or deuterium lamps[161] as a light source, which emits in the near UV and UV regions from 185 nm to 400 nm.

Cuvettes: Cuvettes are small rectangular optical glass or quartz containers, typically with two transparent walls and two opaque walls[162] (Fig. 2-21). Optical glass cuvettes are only suitable for transmission wavelengths in

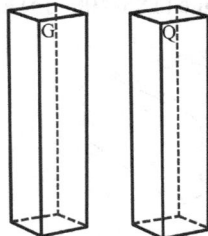

Fig. 2-21　Cuvettes

[151] (溶液)选择性地吸收一定波长的光。溶液呈现的颜色是其强吸收色的互补色……吸光度与样品浓度之间为线性关系

[152] 透光率

[153] 入射光强度

[154] 摩尔吸光系数(或摩尔消光系数)

[155] 比色皿厚度

[156] 单色器

[157] 样品室

[158] 检测器，将辐射能转换为可测量的电信号

[159] 信号处理器和显示器

[160] 钨卤灯

[161] 氢灯或氘灯

[162] 比色皿为矩形的光学玻璃(上方标有 G)或石英(标有 Q)容器，两面透光、两面不透光

the visible spectrum, while quartz cuvettes are generally used for transmission wavelengths in the UV Spectrum. Quartz cuvettes are much more expensive than glass ones, so handle them with care. Optical glass cuvettes are usually labelled with a "G" on one of the upper sides, and quartz cuvettes are labelled with a "Q".

4. How to Use a 722 Visible Spectrophotometer

Ⅰ. Calibration[163]

(1) Turn on the spectrophotometer and warm it up for 20-30 minutes.

(2) Pour the blank (reference) solution into a cuvette and place it in the first position of cuvette shelf[164].

(3) Set the "WAVELENGTH" button to the required wavelength.

(4) Set the "MODE" to "T"; a green light should illuminate.

(5) Pull the bar out to the first position (the light is blocked), press "0%T", and wait for the 0.000 reading to appear on the screen.

(6) Carefully close the cover. Pull the bar back to the end (the light is transmitted) and press "100%T". Wait for the 100.0 reading to appear on the screen.

(7) Repeat the procedure steps (5) and (6), until the reading is stable.

Ⅱ. Measurement

(1) Set the "MODE" to "*A*"; a green light should illuminate.

(2) First, rinse the cuvette with deionized water, then rinse it with the sample solution 2-3 times[165]. Pour the sample solution into the cuvette and place it in the cuvette-shelf. Carefully close the cover.

(3) Wait for the reading to stabilize, then record the absorbance.

Ⅲ. End of the Measurement

(1) Remove all the cuvettes[166]. Turn off the spectrophotometer.

(2) Rinse the cuvettes with deionized water and place them back in the designated container (like a cuvette holder/rack).

5. Notes

(1) Redo the calibration each time the measuring wavelength is changed[167]. At the same wavelength, you can measure different samples after the initial calibration.

(2) Never fill the cuvette more than 2/3 of its volume.

(3) First, dry the cuvette with absorbent paper, then carefully dry both transparent sides with lens cleaning tissue[168].

(4) Never hold the transparent sides of the cuvette; always hold the opaque sides.

[163] 校准

[164] 将空白(参比)溶液倒入比色皿，置于比色皿架的第一格

[165] 用试液润洗比色皿

[166] 取出所有比色皿

[167] 每次测量波长改变时，需重新校准

[168] 先用吸水纸擦干比色皿，再用擦镜纸小心擦干透光面

(5) Dispose of used absorbent paper and waste solutions in the designated containers.

(6) Modern UV-Vis spectrophotometers can automatically perform baseline correction (with a blank solution) and scan the spectrum with a peak-finding function[169]. Watch the videos for more details.

[169] 自动基线校准、光谱扫描并找到最大峰

扫一扫　视频 2-7　分光光度法概述
　　　　视频 2-8　分光光度计的使用

2.5.3　pH Meter

1. Introduction

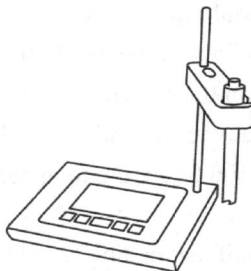

A pH meter (Fig. 2-22) is used to measure the acidity (pH) or alkalinity of a solution[170]. It consists of a voltmeter, an indicator electrode (responsive to H^+)[171], and a reference electrode (with an unchanged potential)[172]. The indicator electrode is usually a glass electrode, while the reference electrode is typically an Ag-AgCl electrode.

[170] pH 计用来测定溶液的酸度(pH)或碱度
[171] 指示电极(响应 H^+)
[172] 参比电极(其电势不变)

Fig. 2-22　pH meter

When these two electrodes are immersed in a solution, they function like a battery. The glass electrode develops a potential that is directly related to the H^+ concentration in the solution, and the voltmeter measures the potential difference between the glass electrode and the reference electrode[173].

Because it is difficult to measure the actual H^+ concentration, pH buffer standards are used to calibration[174]. Commonly, there are three types of pH buffer standards: 4.00, 6.86 and 9.18. The first two are used for acidic solutions, while the latter two are for alkaline solutions.

[173] 玻璃电极产生的电势直接与溶液中 H^+ 浓度有关，电位计测定玻璃电极与参比电极之间的电势差
[174] 用 pH 标准缓冲溶液进行校准

Therefore, the pH of an unknown sample can be expressed by the following equation:

$$pH_x = pH_s + \frac{(E_s - E_x)F}{2.303RT}$$

where E_s is the potential of standard solution, E_x is the potential of unknown sample, T is the thermodynamic temperature (K)[175] of the solution, F is Faraday constant[176] with the value of 96500 C·mol^{-1}.

[175] 热力学温度(K)
[176] 法拉第常量

A glass electrode consists of a glass tube with a small glass bubble at the end[177]. The glass bubble is the active part of the electrode, as it is sensitive to the H^+ concentration of a test sample solution. Typically, the glass electrode and reference electrode are combined into a single unit known as the pH combination electrode[178] (Fig. 2-23).

[177] 底部带有小玻璃泡的玻璃管
[178] 玻璃电极和参比电极组合成 pH 复合电极

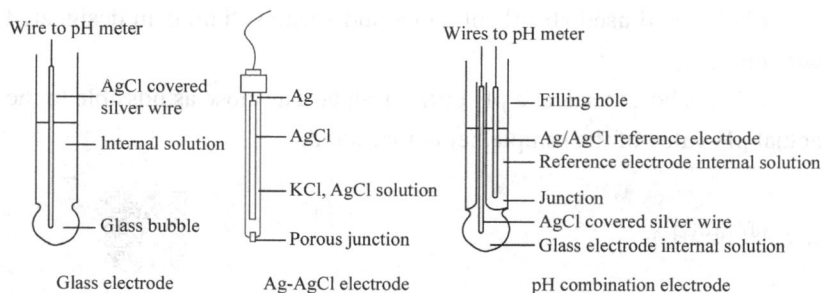

Glass electrode　　　Ag-AgCl electrode　　　pH combination electrode

Fig. 2-23　pH electrodes

2. Operation Guidelines for pH Meter

Ⅰ. Calibration

(1) Press the "Turn on/off" button for several seconds to turn on the pH meter (Fig. 2-24).

(2) Rinse the pH combination electrode with deionized water and dry it with absorbent paper. Immerse the electrode in the first pH buffer standard (pH=6.86)[179], press the "Calibration" button, and stir the solution until the sign of ⌐ on the screen turns to /Ā. Rinse the electrode with deionized water, dry it with absorbent paper, and immerse it in the second pH buffer standard (pH=4.00). Then follow the same procedure.

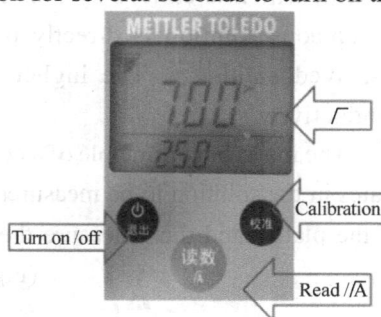

Fig. 2-24　Display of pH meter

[179] 将电极浸入第一种 pH 标准缓冲溶液

(3) Press "Read/⌐" button to finish the calibration procedure[180].

[180] 结束校准过程

Ⅱ. Measurement

Rinse the pH combination electrode with deionized water and dry with absorbent paper. Immerse the electrode in the sample solution, press the "Read//Ā" button, and gently swirl the solution until the ⌐ sign turns to /Ā. Read and record the pH value.

Ⅲ. End of the measurement

(1) Press the "Turn on/off" button for several seconds to turn off the pH meter.

(2) Rinse the pH combination electrode with deionized water and dry it with absorbent paper.

3. Notes

(1) Measure the sample solutions in increasing concentration (from dilute to concentrated)[181]. Clean and dry the pH combination electrode before each measurement.

[181] 按样品浓度增大 (由稀到浓)测定(防止 H+ 的残留)

(2) Discard used absorbent paper and waste solutions in designated containers.

(3) It is best to select a pH buffer standard as close as possible to the actual pH value of the sample being measured.

扫一扫　视频 2-9　酸度计的使用

2.5.4　Conductivity Meter

1. Introduction

Conductivity is the ability of a solution to conduct electrical current, which is carried by ions in the solution. The conductivity of an aqueous solution is directly proportional to the concentration of dissolved solids[182]. The higher the concentration, the greater the conductivity.

The measuring principle of a conductivity meter is to put two parallel plates in the solution to be measured, add a certain potential to both ends of the plates, and then measure the current flow between the plates[183] (Fig. 2-25). Conductance (G) is the reciprocal of resistance (R). The basic unit of conductivity is Siemens (S). L/A is the conductivity cell constant, L is the distance between the two plates, and A is the area of the plate[184].

Fig. 2-25　Conductivity meter

[182] 电导率是溶液中离子传导电流的能力。水溶液的电导率直接与可溶性固体浓度成正比

[183] 将两块平行电板插入待测液中，在电板两端施加电压，再测定电板间的电流

[184] 电导 G 为电阻(R)的倒数。电导率的基本单位为西门子(S)。L/A 为电导池常数，L 为两电板之间的距离，A 为电板面积

$$R = \rho \frac{L}{A} \tag{2-1}$$

$$G = \frac{1}{R} = \kappa \frac{A}{L} \tag{2-2}$$

$$\Lambda_m = \kappa V_m \tag{2-3}$$

$$\Lambda_m = \kappa \frac{1000}{c_m} \quad \text{or} \quad \kappa = \frac{\Lambda_m c_m}{1000} \tag{2-4}$$

$$\Lambda_m^\infty = \sum \lambda_{m,+}^\infty + \sum \lambda_{m,-}^\infty \tag{2-5}$$

Molar conductivity (Λ_m) is the conductivity of a 1 mole electrolyte solution[185]. It can be expressed by Equation (2-4). Limiting molar conductivity (Λ_m^∞) is the conductivity of 1 mole electrolyte at infinite dilution[186]. According to Equation (2-5), Λ_m^∞ is the sum of the limiting molar conductivity of the cations[187] $\Lambda_{m,+}^\infty$ and the anions[188] $\Lambda_{m,-}^\infty$.

[185] 摩尔电导率是 1 mol 电解质溶液的电导率

[186] 极限摩尔电导率是 1 mol 电解质在无限稀释时的电导率

[187] 阳离子

[188] 阴离子

2. Operation Guidelines for a DDS-307 Conductivity Meter

Ⅰ. Calibration

Do not place the electrode in any solution before measurement.

(1) Turn on the conductivity meter and warm up for 20-30 minutes.

(2) Measure the temperature of the solution and calibrate the temperature.

(3) Find the electrode constant, which should be labeled on the electrode[189]. Calibrate the electrode constant.

[189] 电极常数标在电极上

Ⅱ. Measurement

(1) Rinse the electrode with deionized water, then carefully dry it with absorbent paper.

(2) Immerse the electrode in the sample solution.

(3) Read and record the value once it stabilizes.

Ⅲ. Cleaning Up

(1) Turn off the conductivity meter and unplug it.

(2) Rinse the electrode with deionized water and dry with absorbent paper.

3. Notes

(1) Temperature affects the conductivity[190]. To compare measurement results, the test temperature is generally set to 20℃ or 25℃.

[190] 温度影响电导率

(2) Measure the sample solutions in increasing concentration order (from dilute to concentrated)[191]. Clean and dry the electrode before each measurement.

[191] (由稀到浓)测定

(3) Dispose of used absorbent paper and solutions in the designated containers.

2.5.5　Centrifuge

Centrifugation is a technique for separating components, where centrifugal force causes denser particles to move towards the periphery while less dense particles move towards the center[192]. A centrifuge (Fig. 2-26) is used to separate components of a mixture based on their size, density, medium viscosity, and rotor speed[193]. Centrifuges are classified by their maximum speeds, measured in revolutions per minute (rpm)[194]. Speeds range from 0 to 7500 rpm for low-speed centrifuges, and reach 20000 rpm or higher for high-speed centrifuges.

Fig. 2-26　Centrifuge

[192] 离心是一种将混合物分离的技术，离心力使密度大的粒子向外围移动，而密度小的粒子向中心移动

[193] 大小、密度、介质黏度和转子转速

[194] rpm：每分钟转速

Precautions of operating a centrifuge:

(1) Balance the centrifuge: Before starting the centrifuge, it is essential to load it correctly. Balancing the centrifuge prevents potential damage to the instrument and ensures safe operation. For each tube placed in the rotor, insert a tube of equal weight directly opposite it[195] to keep the center of gravity in the rotor. Ensure all sample tubes are evenly filled. If balance tubes[196] are needed, fill them with the same volume of water or a liquid with a similar density to the sample.

(2) Do not open the lid while the rotor is moving, as this is extremely dangerous.

(3) If the centrifuge wobbles or shakes[197], turn it off immediately.

2.5.6　Heating Instruments

There are several heating instruments used in general chemistry laboratories. Some common heating instruments are shown in Fig. 2-27. Always be careful when operating them.

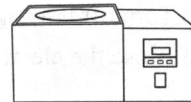

Hot plate[198]　　　Magnetic stirrer hot plate[199]　　　Water bath[200]

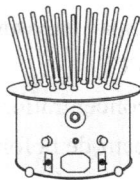

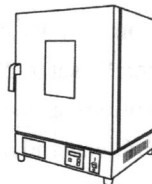

Airflow dryer[201]　　　Oven[202]　　　Muffle furnace[203]

Fig. 2-27　Some common heating instruments

(曾秀琼　蔡吉清编写)

[195] 对于每支插入的离心管，在其正对面加上一支质量相近的离心管
[196] 平衡试管

[197] 摆动或震动

[198] 电炉
[199] 加热磁力搅拌器
[200] 水浴锅

[201] 气流烘干器
[202] 烘箱
[203] 马弗炉

Chapter 3　Fundamental Inorganic Chemistry Experiments
基础无机化学实验

Expt. 1　Separation and Identification of Common Cations
常见阳离子的分离和鉴定

Objectives

(1) Explore the chemical and physical properties of common cations[1].

(2) Understand the principles and procedures for the separation and identification of cations in a mixture[2].

(3) Master general techniques such as reagent addition, centrifugation, and washing precipitates[3].

Principles

Key terms: identification reaction, characteristic reaction, individual analysis, systematic analysis, hydrogen sulfide (H_2S) analysis method, two-acid two-base analysis method.

Inorganic qualitative analysis[4] is the process of separating and identifying inorganic cations or anions[5] in a sample. This process relies on several distinct reactions between ions and reagents, known as identification reactions. These reactions are fast, with high sensitivity and selectivity, and are often accompanied by obvious phenomena such as precipitate formation or dissolution, color change, and gas evolution[6].

When a reagent reacts exclusively with the target ion, producing a unique observable phenomenon without interference[7] from other ions, the reaction is termed a characteristic reaction[8]. However, characteristic reactions are rare, so eliminating interferences to enhance selectivity is crucial. Common methods include pH control, selective precipitation, complexation, and redox reactions[9]. The selective formation of a stable complex with an interfering ion using a chemical reagent to eliminate interferences is called masking[10].

Ions can be directly detected in a test solution using characteristic or identification reactions, regardless of the presence of other ions. This process is known as individual analysis[11]. When characteristic reactions

[1] 阳离子

[2] 混合液

[3] 滴加试剂，离心和洗涤沉淀

[4] 无机定性分析
[5] 阳离子或阴离子
[6] (鉴定反应)反应速度快，具有高灵敏度和高选择性，并常伴随明显的反应现象，如沉淀生成或溶解、颜色变化和气体生成
[7] 干扰
[8] 特征反应

[9] 酸度控制、选择性沉淀、配位反应和氧化还原反应

[10] 掩蔽

[11] 分别分析法

are unavailable or interferences cannot be fully eliminated, ions are first grouped based on similar chemical properties using group reagents[12]. Subsequent separation and identification are then conducted within each group. This process is known as systematic analysis[13]. Common methods for systematic analysis of cations include the hydrogen sulfide (H_2S) analysis method[14] (Fig. 3-1) and the two-acid two-base analysis method[15] (Fig. 3-2).

[12] 组试剂

[13] 系统分析法

[14] 硫化氢(H_2S)分析法
[15] 两酸两碱分析法

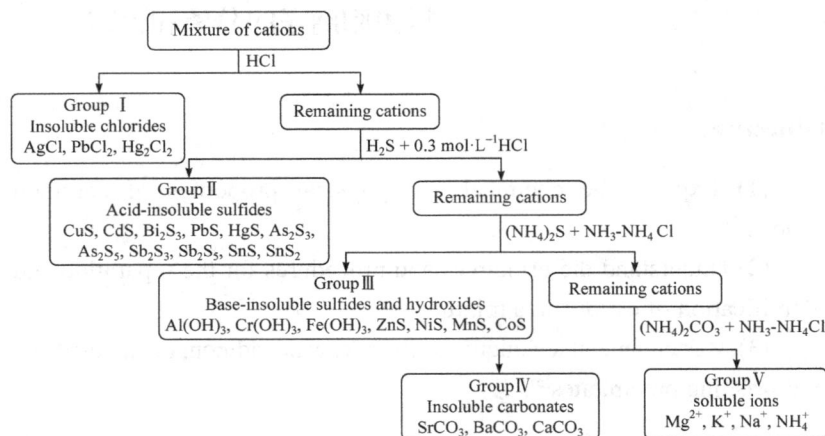

Fig. 3-1 Hydrogen sulfide (H_2S) analysis method
Note: $PbCl_2$ precipitates at high concentration.

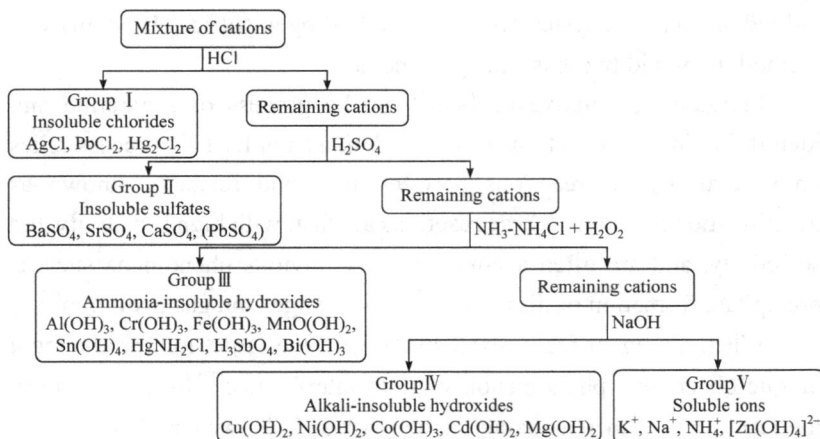

Fig. 3-2 Two-acid two-base analysis method
Note: (1) $PbCl_2$ or $CaSO_4$ precipitates at high concentrations. (2) $Cu(OH)_2$ dissolves in excess strong alkali.

In this experiment, common cations such as Ag^+, Pb^{2+}, Cu^{2+}, Fe^{3+}, Fe^{2+}, Al^{3+}, Cr^{3+}, Mn^{2+}, Co^{2+}, Ni^{2+}, and Zn^{2+} will be separated and identified using the two-acid two-base analysis method. During the tests, special attention should be given to factors such as solvent, temperature, catalyst, concentration, and the acidity or basicity of the solution, as these can influence the identification reaction[16].

[16] 特别注意溶剂、温度、催化剂、浓度和酸碱度等因素对鉴定反应的影响

Pre-lab Questions

(1) Among the following cations, Ag^+, Pb^{2+}, Cu^{2+}, Fe^{3+}, Fe^{2+}, Al^{3+}, Cr^{3+}, Mn^{2+}, Co^{2+}, Ni^{2+}, and Zn^{2+}:

1) Which form insoluble chlorides[17]?　　　　　　　　　　[17] 不溶的氯化物

2) Which form insoluble sulfates[18]?　　　　　　　　　　[18] 不溶的硫酸盐

3) Which react with sodium hydroxide (NaOH) to form precipitates[19]?　[19] 沉淀
Among these, which precipitates can dissolve in excess NaOH solution?

4) Which react with ammonia water[20] ($NH_3 \cdot H_2O$) to form　[20] 氨水
precipitates? Among these, which precipitates can dissolve in excess $NH_3 \cdot H_2O$?

(2) How can we verify the precipitation of a cation is complete?

Apparatus and Reagents

Apparatus: centrifuge[21], spot plate[22] (black, white), centrifuge　[21] 离心机
tube[23], test tube[24].　　　　　　　　　　　　　　　　　　　[22] 点滴板

Reagents: $NaBiO_3$ (s), HNO_3 (6 mol·L^{-1}, 2 mol·L^{-1}), HCl (6 mol·L^{-1},　[23] 离心管
2 mol·L^{-1}), H_2SO_4 (3 mol·L^{-1}, 2 mol·L^{-1}), HAc (6 mol·L^{-1}, 2 mol·L^{-1}),　[24] 试管
NaOH (6 mol·L^{-1}, 2 mol·L^{-1}), $NH_3 \cdot H_2O$ (6 mol·L^{-1}, 2 mol·L^{-1}), alizarin
red S (0.1%), saturated NH_4F, saturated NH_4SCN, H_2O_2 (3%), 1,10-
phenanthroline (1%), dimethylglyoxime (1%), dithizone-CCl_4 solution
(0.01%), diethyl ether, acetone, pH test paper (1-14).

The concentrations of following solutions are 0.1 mol·L^{-1}: $AgNO_3$,
$Pb(NO_3)_2$, K_2CrO_4, $CuSO_4$, $FeCl_3$, $FeSO_4$, $K_4[Fe(CN)_6]$, $K_3[Fe(CN)_6]$,
$KSCN$, $CoCl_2$, $AlCl_3$, $CrCl_3$, $MnSO_4$, $NiCl_2$, $ZnSO_4$.

The concentration of each cation in the mixture is 0.1 mol·L^{-1}.

Procedures

1. Individual Confirmatory Tests for Common Cations[25]

The procedures of individual confirmatory tests for common cations,　[25] 常见阳离子的单个
such as Ag^+, Pb^{2+}, Cu^{2+}, Fe^{3+}, Fe^{2+}, Al^{3+}, Cr^{3+}, Mn^{2+}, Co^{2+}, Ni^{2+}, and Zn^{2+},　鉴定实验
are listed in Table 3-1.

Table 3-1　The procedures of individual confirmatory tests for common cations

Cations	Procedures a)
$Ag^{+\,b)}$	(1) 2 drops of 0.1 mol·L^{-1} $AgNO_3$ + 1 drop of 2 mol·L^{-1} HCl (2) Centrifuge and discard the supernatant[26]. (3) Precipitate + 6 mol·L^{-1} $NH_3 \cdot H_2O$ dropwise (4) Solution obtained from step (3) + 6 mol·L^{-1} HNO_3
Pb^{2+}	5 drops of 0.1 mol·L^{-1} $Pb(NO_3)_2$ + 1 drop of 6 mol·L^{-1} HAc + 5 drops of 0.1 mol·L^{-1} K_2CrO_4

[26] 离心并弃去上层清液

Continued

Cations	Procedures a)
Cu^{2+} c)	3 drops of 0.1 mol·L^{-1} CuSO$_4$ + 1 drop of 2 mol·L^{-1} HAc + 1 drop of 0.1 mol·L^{-1} K$_4$[Fe(CN)$_6$]
Fe^{3+} d)	Method 1: 1 drop of 0.1 mol·L^{-1} FeCl$_3$ + 1 drop of 0.1 mol·L^{-1} K$_4$[Fe(CN)$_6$] Method 2: 1 drop of 0.1 mol·L^{-1} FeCl$_3$ + 2 drops of saturated[27] NH$_4$SCN
Fe^{2+}	Method 1: 5 drops of 0.1 mol·L^{-1} FeSO$_4$ + 5 drops of 0.1 mol·L^{-1} K$_3$[Fe(CN)$_6$] Method 2: 10 drops of 0.1 mol·L^{-1} FeSO$_4$ + 5 drops of 1% 1,10-phenanthroline solution in ethanol[28]
Al^{3+} e)	1 drop of 0.1 mol·L^{-1} AlCl$_3$ + 1 drop of 2 mol·L^{-1} HAc + 2 drops of 0.1% alizarin red S[29] + 2 mol·L^{-1} NH$_3$·H$_2$O dropwise to weakly basic + heat in a water bath[30]
Cr^{3+}	(1) 5 drops of 0.1 mol·L^{-1} CrCl$_3$ + 6 mol·L^{-1} NaOH dropwise until the greyish-green precipitate dissolves and a bright green solution is formed[31] (2) Solution obtained from step (1) + 6-7 drops of 3% H$_2$O$_2$ + heat in a water bath (3) A portion of the solution obtained from step (2) + 6 mol·L^{-1} HNO$_3$ + 2 drops of 0.1 mol·L^{-1} Pb(NO$_3$)$_2$ (4) A portion of the solution obtained from step (2) + 6 mol·L^{-1} HNO$_3$ until the pH reaches 2-3 + half a dropper of diethyl ether[32] + 2 mL of 3% H$_2$O$_2$ + shake the tube
Mn^{2+}	2 drops of 0.1 mol·L^{-1} MnSO$_4$ + 10 drops of 6 mol·L^{-1} HNO$_3$ + a small amount of NaBiO$_3$ + heat in a water bath
Co^{2+} f)	5-6 drops of 0.1 mol·L^{-1} CoCl$_2$ + 2 drops of 2 mol·L^{-1} HCl + 5-6 drops of saturated NH$_4$SCN + 10 drops of acetone[33] + shake the tube
Ni^{2+} g)	1 drop of 0.1 mol·L^{-1} NiCl$_2$ + 1 drop of 2 mol·L^{-1} NH$_3$·H$_2$O + 1 drop of 1% dimethylglyoxime[34]
Zn^{2+}	3 drops of 0.1 mol·L^{-1} ZnSO$_4$ + 6-7 drops of 2 mol·L^{-1} NaOH + half a dropper of 0.01% dithizone[35]-CCl$_4$ + shake the tube

Note:

a) The identification of these anions is valid in the absence of other interfering ions.

b) Test Ag$^+$ in centrifuge tubes.

c) Fe^{3+} interferes with the identification and can be masked with NH$_4$F.

d) Test Fe^{3+} on a white spot plate[36].

e) Al^{3+} reacts with alizarin red S in HAc-NaAc buffer (pH 4-5) to form a red complex. A weakly basic system and heating can promote the formation of bright red flocculent precipitates.

f) Fe^{3+} or a significant amount of Cu^{2+} interferes with the identification. Mask Fe^{3+} with saturated NH$_4$F, and reduce Cu^{2+} with Na$_2$SO$_3$.

g) Test Ni^{2+} on a white spot plate. Fe^{2+} reacts with dimethylglyoxime to form a red soluble chelate in NH$_3$·H$_2$O[37]. To eliminate the interference of Fe^{2+}, oxidize it to Fe^{3+} with H$_2$O$_2$, then mask with citric acid or tartaric acid[38].

[27] 饱和的

[28] 1%邻二氮菲乙醇溶液

[29] 茜素磺酸钠

[30] 水浴加热

[31] 灰绿色沉淀溶解，形成亮绿色溶液

[32] 半滴管乙醚

[33] 丙酮

[34] 丁二酮肟

[35] 二苯硫腙

[36] 白色点滴板

[37] Fe^{2+}和丁二酮肟在氨水中形成红色可溶性的螯合物

[38] 用柠檬酸或酒石酸掩蔽

[39] 已知阳离子混合液的分析

[40] 将上层清液倒入干净的试管中

[41] 酸化

[42] 丁二酮肟

2. Analysis of Known Mixture of Cations[39]

Ⅰ. Separation and Identification of Mixed Fe^{3+}, Cr^{3+}, Mn^{2+} and Ni^{2+}

The procedures of separation and identification of mixed Fe^{3+}, Cr^{3+}, Mn^{2+} and Ni^{2+} are shown in Fig. 3-3.

(1) Add 10 drops of the mixed solution and 10 drops of 2 mol·L^{-1} NH$_3$·H$_2$O to a centrifuge tube. Stir well, centrifuge, and decant the supernatant into a clean test tube[40]. Save the precipitate for further testing.

On a white spot plate, add 2 drops of the supernatant. Acidify[41] with 2-3 drops of 6 mol·L^{-1} HAc, then add 2 drops of 1% dimethylglyoxime[42].

$Fe^{3+}, Cr^{3+}, Mn^{2+}, Ni^{2+}$

$2\ mol\cdot L^{-1}\ NH_3\cdot H_2O$

$Fe(OH)_3\downarrow, Cr(OH)_3\downarrow, Mn(OH)_2\downarrow(MnO_2\downarrow)$ | $[Ni(NH_3)_6]^{2+}$

$2\ mol\cdot L^{-1}\ NaOH$
$3\%\ H_2O_2$
Heat

$6\ mol\cdot L^{-1}\ HAc$
$1\%\ dimethylglyoxime$

Bright red precipitate confirms Ni^{2+}

$Fe(OH)_3\downarrow, MnO_2\downarrow$ | CrO_4^{2-}

$2\ mol\cdot L^{-1}\ H_2SO_4$

$6\ mol\cdot L^{-1}\ HAc$
$0.1\ mol\cdot L^{-1}\ Pb(NO_3)_2$

$MnO_2\downarrow$ | Fe^{3+}

$3\ mol\cdot L^{-1}\ H_2SO_4$
$3\%\ H_2O_2$
Heat

$0.1\ mol\cdot L^{-1}\ KSCN$

Dark red solution confirms Fe^{3+}

$PbCrO_4\downarrow$
Yellow precipitate confirms Cr^{3+}

Mn^{2+}

$NaBiO_3(s)$

MnO_4^-
Purplish red solution confirms Mn^{2+}

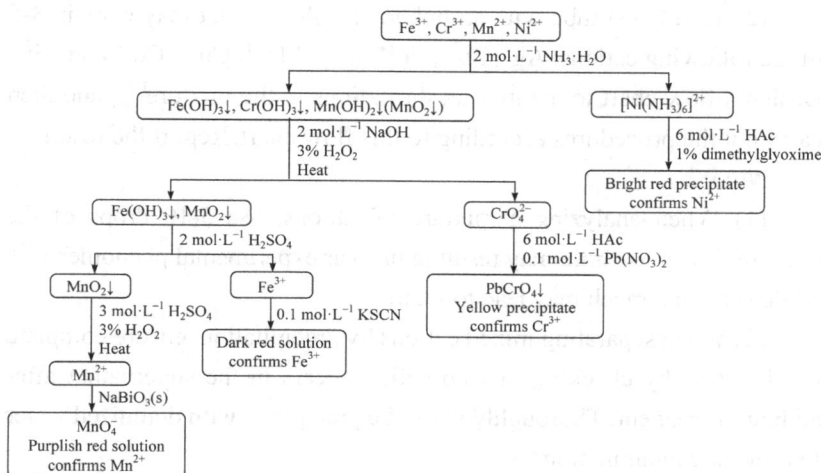

Fig. 3-3　Separation and identification of mixed Fe^{3+}, Cr^{3+}, Mn^{2+} and Ni^{2+}

Record the observations to identify Ni^{2+}.

(2) Add 8 drops of $2\ mol\cdot L^{-1}$ NaOH and 10 drops of 3% H_2O_2 to the precipitate from step (1). Heat in a water bath, then centrifuge. Decant the supernatant into a clean test tube and keep the precipitate for further testing.

Transfer 5 drops of the supernatant into a clean test tube. Acidify with 2-3 drops of $6\ mol\cdot L^{-1}$ HAc, then add 2 drops of $0.1\ mol\cdot L^{-1}$ $Pb(NO_3)_2$. Record the observations to identify Cr^{3+}.

(3) Wash the precipitate from step (2) with deionized water. Centrifuge and discard the supernatant. Add 3 drops of $2\ mol\cdot L^{-1}$ H_2SO_4 to the precipitate. Heat in a water bath, then centrifuge. Decant the supernatant into a clean test tube and keep the precipitate for further testing.

Add 1 drop of $0.1\ mol\cdot L^{-1}$ KSCN to the supernatant. Record the observations to identify Fe^{3+}.

(4) Add 2 drops of $3\ mol\cdot L^{-1}$ H_2SO_4 and 3 drops of 3% H_2O_2 to the precipitate from step (3). Heat in a boiling water bath to remove excess H_2O_2[43], and dissolve the precipitate completely. Add 4 drops of $3\ mol\cdot L^{-1}$ H_2SO_4 and a small amount of $NaBiO_3$. Record the observations to identify Mn^{2+}.

II. Separation and Identification of Mixed Ag^+, Fe^{3+}, Al^{3+} and Cu^{2+}

Design a flow chart that represents the above before class and then conduct the tests during class[44].

3. Analysis of Unknown Mixture of Cations[45]

(1) Take a portion of unknown solution I or II, which may contain 2-3 of the following cations: Fe^{3+}, Ni^{2+}, Cr^{3+} and Mn^{2+}. Follow the procedure outlined in Fig. 3-3 to identify the cations present. Report the results.

[43] 沸水浴加热以除去过量的过氧化氢

[44] 课前设计相关的流程图，课中完成试验

[45] 未知阳离子混合液的分析

[46] 设计鉴定该混合液中所有阳离子的流程图

(2) Take a test tube with an unknown solution that may contain 3-5 of the following cations: Ag^+, Pb^{2+}, Fe^{3+}, Cr^{3+}, Mn^{2+}, Cu^{2+}, Co^{2+} and Ni^{2+}. Design a flow chart to identify all the cations in the mixture[46], and then carry out the procedures according to this flow chart. Report the results.

Note:

[47] 过少可能导致实验现象不明显

(1) When analyzing a mixture of cations, use 5-10 drops of the mixture. Using too little may result in unclear experimental phenomena[47], while using too much may lead to waste.

(2) When separating mixed cations by precipitation, ensure complete precipitation by checking that no solid appears in the supernatant after adding the reagent. Thoroughly wash the precipitate with deionized water to prevent contamination[48].

[48] 避免污染

Safety and Waste Disposal

(1) Wear a lab coat and safety goggles at all times in the laboratory. Wear gloves when handling hot glassware. In case of chemical splashes into the eyes or onto the skin, immediately rinse the affected area with plenty of cold water.

(2) Verify the label on all containers before using any chemicals.

(3) Never point the test tube at yourself or others when heating or adding reagents.

[49] CCl₄、丙酮和乙醚等有机溶剂易挥发且危险。在通风橱内操作

(4) Organic solvents such as CCl_4, acetone and diethyl ether are volatile and hazardous. Handle them in a fume hood[49].

[50] 确保将等重的离心管对称放置在对角位置使其平衡

(5) Handle the centrifuge with care. Ensure it is balanced by placing equal-weighted tubes diagonally opposite each other[50]. Do not open the lid until the centrifuge has come to a complete stop.

(6) Discard all waste into the designated waste container.

Post-lab Questions

(1) In the separation and identification of mixed Fe^{3+}, Cr^{3+}, Mn^{2+} and Ni^{2+}, a student failed to detect Mn^{2+}. Provide a brief explanation.

(2) How can Ag^+ and Pb^{2+} be separated and identified?

(3) In Procedures 2.1, can Fe^{3+}, Cr^{3+}, Mn^{2+}, and Ni^{2+} be separated and identified if sodium hydroxide (NaOH) is used instead of an ammonia water ($NH_3 \cdot H_2O$) in the first step? Provide a brief explanation.

Exploring Experiments

Iron is a crucial element for human nutrition and is commonly included in dietary supplements to address iron deficiency. It exists in two valence states: ferrous (Fe^{2+}) and ferric (Fe^{3+}). These forms differ in solubility and absorption: Fe^{2+} is more soluble and better absorbed, whereas Fe^{3+} often forms insoluble compounds at neutral pH. Design a procedure to determine the valence state of iron in various commercially available oral iron supplements.

扫一扫　视频 3-1　常见阳离子的分离和鉴定(讲解)
　　　　视频 3-2　常见阳离子的单个鉴定(实验)

扫一扫　视频 3-3　已知阳离子混合液的分离和鉴定(实验)
　　　　视频 3-4　离心、沉淀洗涤和滴瓶使用(实验)

(徐孝菲编写)

Expt. 2　Separation and Identification of Common Anions
常见阴离子的分离和鉴定

Objectives

(1) Explore the chemical and physical properties of common anions[51].

(2) Understand the principles and procedures for the separation and identification of anions in a mixture.

(3) Master general techniques such as reagent addition, centrifugation, and washing precipitates[52].

Principles

Key terms: individual analysis, characteristic reactions, preliminary tests.

Common anions are simple ions or oxoacid ions[53] composed of non-metallic elements[54], such as X^- (X=F, Cl, Br, I), S^{2-}, SO_4^{2-} and ClO_3^-. Most anions exhibit minimal mutual interference, with only a few capable of coexisting[55]. Many anions exhibit unique characteristic reactions[56] and are prone to transformation during analysis. Therefore, individual analysis[57] is typically used for anions instead of complex systematic analysis[58]. When interfering ions are known, appropriate separation or masking[59] techniques can be applied.

[51] 阴离子

[52] 滴加试剂，离心和洗涤沉淀

[53] 含氧酸根离子
[54] 非金属元素
[55] 大多数阴离子相互干扰较小，只有很少能共存
[56] 特征反应
[57] 分别分析法
[58] 系统分析法
[59] 掩蔽

[60] 初步试验

When analyzing a mixture of anions, preliminary tests[60] are conducted based on their analytical properties. Once possible anions are identified, individual analyses are performed. The analytical characteristics of anions are primarily as follows:

[61] 低沸点或易分解酸

(1) Anions from low-boiling or easily decomposed acids[61] can react with dilute acids, releasing various gases. Analyzing the properties of these gases provides preliminary information about the presence of anions such as CO_3^{2-}, S^{2-}, SO_3^{2-}, $S_2O_3^{2-}$ and NO_2^-. Table 3-2 summarizes the physical and chemical properties of these gases.

Table 3-2　Properties of gases evolved by anion-acid reactions

Observations			Inferences	
Color of gases	Smell of gases	Characteristics of gases	Gas evolved	Possible anions
colorless	odourless	turns lime water turbid	CO_2 a)	CO_3^{2-}
colorless	pungent smell like burning sulphur	fades the color of I_2- starch solution or $KMnO_4$ solution	SO_2 b)	SO_3^{2-}, $S_2O_3^{2-}$
colorless	rotten eggs smell	turns moistened $PbAc_2$ test paper black	H_2S	S^{2-}
brown	pungent odour	—	NO, NO_2	NO_2^-

Note:
a) SO_2 also turns lime water turbid.
b) H_2S also fades the color of I_2-starch solution or dilute $KMnO_4$ solution.

[62] 碱金属盐

(2) Except for salts of alkali metals[62] and those formed by NO_3^-, ClO_3^-, ClO_4^- or Ac^-, most other salts are largely insoluble. The 15 common anions are classified into three groups based on the solubility of

[63] 钡盐和银盐

their barium (Ba) salts and silver (Ag) salts[63] (Table 3-3).

Table 3-3　Grouping of 15 common anions

Group	Group reagents	Anions	Characteristics
Group I	$BaCl_2$ (neutral or weakly basic medium)	CO_3^{2-}, SO_4^{2-}, SO_3^{2-}, $S_2O_3^{2-}$, SiO_3^{2-}, PO_4^{3-}, AsO_3^{3-}, AsO_4^{3-} a)	Barium salts are insoluble in water b), and silver salts are soluble in HNO_3
Group II	$AgNO_3$ (cold dilute HNO_3)	Cl^-, Br^-, I^-, S^{2-}	Silver salts are insoluble in water or dilute HNO_3 c)
Group III	—	NO_2^-, NO_3^-, Ac^-	Barium salts and silver salts are both soluble in water

Note:
a) Precipitation occurs at high concentration.
b) Barium salts are soluble in acid except $BaSO_4$.
c) Ag_2S is soluble in hot HNO_3.

[64] 氧化性或还原性

(3) Most anions exhibit oxidizing or reducing properties[64], with exceptions such as Ac^-, CO_3^{2-}, SO_4^{2-} and PO_4^{3-}. These anions may react

with each other. In an acidic medium, anions with strong reducing properties, such as S^{2-}, SO_3^{2-} and $S_2O_3^{2-}$, can be oxidized by I_2. A color fade[65] after adding I_2-starch solution[66] indicates their presence. Treatment with the strong oxidant $KMnO_4$ can reveal even weakly reducing anions, such as Br^- and I^-. If the reddish-purple color fades, reducing anions are present; if not, they are absent. Cl^- being a weaker reductant[67] compared to Br^- and I^-, can only reduce $KMnO_4$ at high concentrations of Cl^- and H^+.

In an acidic medium, the oxidizing anion NO_2^- can oxidize I^- to I_2, resulting in a blue color in a starch solution or a reddish-purple color in CCl_4. The oxidizability[68] of NO_3^- is weak, so such reactions occur only at high concentrations. The redox reaction between AsO_4^{3-} and I^- is reversible. In neutral or weakly basic conditions, I_2 oxidizes AsO_3^{3-} to AsO_4^{3-}.

Possible anions can be inferred through preliminary tests based on the aforementioned analytical characteristics. Subsequent separation and identification of individual anions are performed according to their distinct properties and characteristic reactions. Preliminary tests for 15 common anions are summarized in Table 3-4.

[65] 褪色

[66] 碘-淀粉溶液

[67] 还原剂

[68] 氧化能力

Table 3-4　Preliminary tests for 15 common anions

Anions	H_2SO_4	$BaCl_2$ (neutral or basic media)	$AgNO_3$ (dilute HNO_3)	I_2-starch (dilute H_2SO_4)	$KMnO_4$ (dilute H_2SO_4)	KI-starch (dilute H_2SO_4)
SO_4^{2-}		+				
SO_3^{2-}	+	+		+	+	
$S_2O_3^{2-}$	+	(+)	+	+	+	
CO_3^{2-}	+	+				
PO_4^{3-}		+				
AsO_4^{3-}		+				+
AsO_3^{3-}		(+)			+	
SiO_3^{2-}	(+)	+				
Cl^-			+		(+)	
Br^-			+		+	
I^-			+		+	
S^{2-}			+	+	+	
NO_2^-					+	+
NO_3^-						(+)
Ac^-						

Note: "+" indicates that reactions occur (positive result). "(+)" signifies that reactions occur only at high concentrations.

Pre-lab Questions

[69] 共存

(1) Among the 15 anions listed in Table 3-4, which can coexist[69]? Which can react with each other? Give the reaction equations.

[70] 稀的

[71] 浓的

(2) In redox experiments, can dilute[70] HNO_3, dilute HCl, or concentrated[71] H_2SO_4 be used instead of dilute H_2SO_4 to acidify the solution? Give the reason.

[72] 中性溶液

(3) Which anions might be present in a neutral solution[72] that contains Ba^{2+} and Ag^+?

Apparatus and Reagents

[73] 离心机

[74] 点滴板

Apparatus: centrifuge[73], centrifuge tube, test tube, spot plate[74].

Reagents: $PbCO_3$ (s), $NaNO_2$ (s), $FeSO_4$ (s), Zn (s), Pb(Ac)$_2$ test paper, H_2SO_4 (concentrated, 3 $mol·L^{-1}$, 1 $mol·L^{-1}$), HNO_3 (6 $mol·L^{-1}$, 2 $mol·L^{-1}$), diphenylamine-concentrated H_2SO_4 solution, HCl (6 $mol·L^{-1}$, 2 $mol·L^{-1}$), $NH_3·H_2O$ (2 $mol·L^{-1}$), $BaCl_2$ (0.5 $mol·L^{-1}$), $SrCl_2$ (0.5 $mol·L^{-1}$), I_2 (0.5 $mol·L^{-1}$), $Na_2[Fe(CN)_5NO]$ (3%), CCl_4, starch (0.2%).

The concentrations of following solutions are 0.1 $mol·L^{-1}$: Na_2S, Na_2SO_3, $Na_2S_2O_3$, Na_2SO_4, $NaNO_2$, Na_3PO_4, NaCl, NaClO, KBr, KI, $NaNO_2$, $AgNO_3$, $(NH_4)_2MoO_4$.

The concentration of each anion in the mixture is 0.1 $mol·L^{-1}$.

Procedures

[75] 常见阴离子的单个鉴定实验

1. Individual Confirmatory Tests for Common Anions[75]

The procedures of individual confirmatory tests for some common anions are listed in Table 3-5.

Table 3-5   The procedures of individual confirmatory tests for common anions

Anions	Procedures a)
Cl^-	(1) 5 drops of 0.1 $mol·L^{-1}$ NaCl + 1 drops of 2 $mol·L^{-1}$ HNO_3 + 5 drops of 0.1 $mol·L^{-1}$ $AgNO_3$ (2) Precipitate obtained from step (1) + excess[76] 6 $mol·L^{-1}$ $NH_3·H_2O$ (3) Solution obtained from step (2) + excess 6 $mol·L^{-1}$ HNO_3
Br^- or I^-	5 drops of 0.1 $mol·L^{-1}$ KBr or 0.1 $mol·L^{-1}$ KI + 2-4 drops of 3 $mol·L^{-1}$ H_2SO_4 + 10 drops of CCl_4 + NaClO solution + shake the test tube
NO_2^-	5 drops of 0.1 $mol·L^{-1}$ $NaNO_2$ + 2 drops of 3 $mol·L^{-1}$ H_2SO_4 + 3 drops of 0.1 $mol·L^{-1}$ KI + 10 drops of CCl_4 + shake the test tube
NO_3^-	Method 1: 5 drops of 0.1 $mol·L^{-1}$ $NaNO_3$ + 2 drops of 1 $mol·L^{-1}$ H_2SO_4 + half a dropper of diphenylamine-concentrated H_2SO_4 solution[77] (added along the side of the test tube[78]) Method 2 a): 1 drop of 0.1 $mol·L^{-1}$ $NaNO_3$ + a grain of $FeSO_4$ crystal + 3-4 drops of concentrated H_2SO_4[79] (added along the edge of the crystal[80])
$S_2O_3^{2-}$	Method 1: 3 drops of 0.1 $mol·L^{-1}$ $Na_2S_2O_3$ + 5 drops of 0.1 $mol·L^{-1}$ $AgNO_3$ Method 2: 5 drops of 0.1 $mol·L^{-1}$ $Na_2S_2O_3$ + 2 drops of 2 $mol·L^{-1}$ HCl + heat in a water bath[81]

[76] 过量的

[77] 半滴管二苯胺-浓硫酸溶液

[78] 沿试管壁滴加

[79] 浓硫酸

[80] 沿晶体边缘滴加

[81] 水浴加热

Continued

Anions	Procedures [a)
SO_3^{2-}	(1) 5 drops of 0.1 mol·L^{-1} Na$_2$SO$_3$ + 3 drops of I$_2$ -starch solution (2) Solution obtained from step (1) + 2 drops of 2 mol·L^{-1} HCl
S^{2-}	Method 1: 5 drops of 0.1 mol·L^{-1} Na$_2$S + 2 drops of 2 mol·L^{-1} HCl + a piece of moistened PbAc$_2$ test paper[82] above the mouth of the test tube Method 2[b)]: 1 drop of 0.1 mol·L^{-1} Na$_2$S + 1 drop of 3% Na$_2$[Fe(CN)$_5$NO]
SO_4^{2-}	(1) 5 drops of 0.1 mol·L^{-1} Na$_2$SO$_4$ + 1 drop of 0.5 mol·L^{-1} BaCl$_2$ (2) Precipitate obtained in step (1) + 5 drops of 2 mol·L^{-1} HCl
PO_4^{3-}	3 drops of 0.1 mol·L^{-1} Na$_3$PO$_4$ + 6 drops of 6 mol·L^{-1} HNO$_3$ + 10 drops of 0.1 mol·L^{-1} (NH$_4$)$_2$MoO$_4$ + heat in a water bath

Note:

a) The identification of these anions is valid in the absence of other interfering ions.

b) Test NO$_3^-$ or S^{2-} using Method 2 on a white spot plate[83].

[82] 湿润的乙酸铅试纸

[83] 白色点滴板

2. Analysis of Known Mixture of Anion[84]

[84] 已知阴离子混合液的分析

Ⅰ. Separation and Identification of Mixed S^{2-}, SO_3^{2-} and $S_2O_3^{2-}$

The flow chart of separation and identification of mixed S^{2-}, SO_3^{2-} and $S_2O_3^{2-}$ are shown in Fig. 3-4.

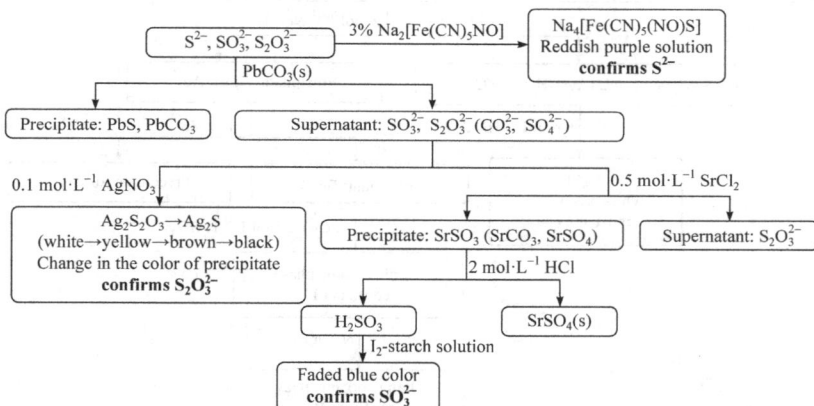

Fig. 3-4　Separation and identification of mixed S^{2-}, SO_3^{2-} and $S_2O_3^{2-}$

(1) Add 1 drop of the mixed solution onto a spot plate. Add 1 drop of Na$_2$[Fe(CN)$_5$NO] solution. The appearance of a reddish-purple color confirms the presence of S^{2-}.

(2) To remove the interference[85] of S^{2-}, add a small amount of PbCO$_3$ to 10 drops of the mixed solution. Stir well and centrifuge. Decant the supernatant into a clean test tube and discard the precipitate[86]. Transfer 1 drop of the supernatant into a test tube, add 1 drop of Na$_2$[Fe(CN)$_5$NO] solution, and observe for any color change. If no color change is observed, proceed with the subsequent experiments.

[85] 干扰

[86] 将上清液倾析至干净试管中，并弃去沉淀

(3) Transfer 1 drop of the supernatant without S^{2-} onto a spot plate. Add 2 drops of 0.1 mol·L^{-1} AgNO$_3$. Record observations to determine the presence of $S_2O_3^{2-}$.

[87] 若沉淀未完全溶解，
则离心弃去固体

(4) Add 4-5 drops of the supernatant without S^{2-} into a centrifuge tube. Add 0.5 mol·L^{-1} SrCl$_2$ dropwise until precipitation is complete. Heat in a water bath for 3 minutes, then cool and centrifuge. Decant and discard the supernatant. Wash the precipitate with water, and centrifuge again. Discard the supernatant. Add 3-4 drops of 2 mol·L^{-1} HCl to the precipitate and stir well. If the precipitate does not dissolve completely, centrifuge and discard the solid[87]. Add 1 drop of I$_2$-starch solution to the supernatant. Record observations to determine the presence of SO_3^{2-}.

Ⅱ. Separation and Identification of Mixed Cl⁻, Br⁻ and I⁻

The procedures of separation and identification of mixed Cl⁻, Br⁻ and I⁻ are shown in Fig. 3-5.

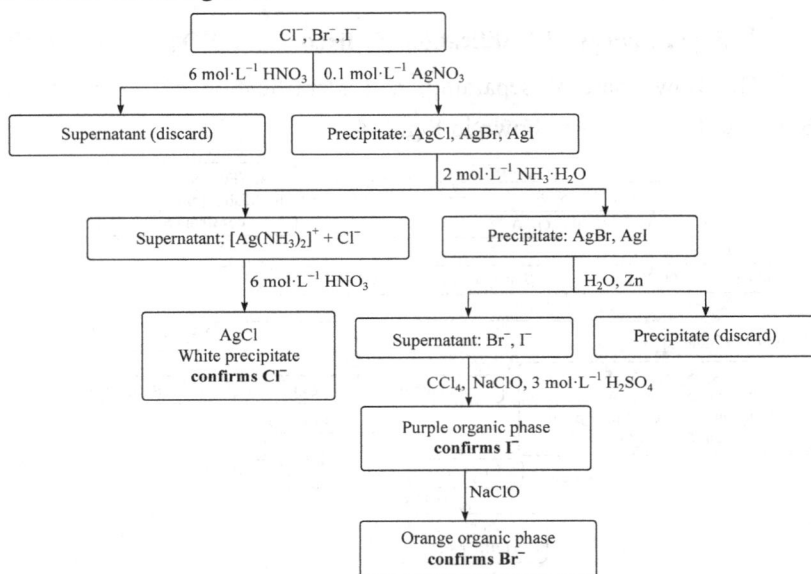

Fig. 3-5　Separation and identification of mixed Cl⁻, Br⁻ and I⁻

(1) Acidify 1 mL of mixed solution with 2 drops of 6 mol·L^{-1} HNO$_3$. Add 0.1 mol·L^{-1} AgNO$_3$ dropwise until precipitation is complete. Heat the mixture in a water bath for 2 minutes. Centrifuge, discard the supernatant, and wash the precipitate with water twice. Discard the supernatant and save the precipitate for further testing[88].

[88] 弃去上层清液，保留
沉淀做下一步实验

[89] 酸化

(2) Add 1 mL of 2 mol·L^{-1} NH$_3$·H$_2$O to the precipitate obtained. Stir for 1 minute and centrifuge. Wash the precipitate with water and save it for further testing. Acidify[89] the supernatant with 6 mol·L^{-1} HNO$_3$. Record observations to determine the presence of Cl⁻.

(3) Add 6 drops of water and appropriate amount of Zn to the precipitate obtained above. Stir for 2-3 minutes and centrifuge. Decant the supernatant into a clean test tube and discard the precipitate.

Add 4-5 drops of CC1$_4$ to the supernatant. Add 4-5 drops of 3 mol·L^{-1} H$_2$SO$_4$ and 1 drop of freshly prepared NaClO solution[90], and stir well. Record observations to determine the presence of I$^-$. Next, add another drop of freshly prepared NaClO solution, and stir well. Record observations to determine the presence of Br$^-$.

[90] 新配制的次氯酸钠溶液

Safety and Waste Disposal

(1) Wear a lab coat and safety goggles at all times in the laboratory. Wear gloves when handling hot glassware. In case of chemical splashes into the eyes or onto the skin, immediately rinse the affected area with plenty of cold water.

(2) Verify the label on all containers before using any chemicals. When smelling fumes, gently waft the vapors towards your nose[91].

[91] 嗅闻气体时，轻轻将气体扇动至鼻子附近

(3) Never point the test tube at yourself or others when heating or adding reagents.

(4) Handle the centrifuge with care. Ensure it is balanced by placing equal-weighted tubes diagonally opposite each other[92]. Do not open the lid until the centrifuge has come to a complete stop.

[92] 确保将等重的离心管对称放置在对角位置使其平衡

(5) Discard all waste into the designated waste containers.

Post-lab Questions

(1) In the separation and identification of Br$^-$ and I$^-$, what is the purpose of adding CCl$_4$? Does it participate in chemical reactions?

(2) Are there interferences between S^{2-}, SO$_3^{2-}$ and S$_2$O$_3^{2-}$ during their identification? How can these interferences be removed?

(3) In a neutral solution[93] of unknown anions, the addition of dilute H$_2$SO$_4$ produces bubbles. Testing with barium salts (Ba^{2+}) or silver salts (Ag$^+$) yields negative results, while tests with KMnO$_4$ or KI-starch test paper show positive results. Which anions might be present, and which anions are difficult to confirm?

[93] 中性溶液

Exploring Experiments

According to news reports, some community water purifiers[94] may not provide fully purified water[95], raising concerns about its quality and safety. Tap water contains ions such as chloride (Cl$^-$) and sulfate (SO$_4^{2-}$), commonly introduced during water treatment. Assessing the removal of these ions, particularly chloride, can indicate the purifier's effectiveness. Design a procedure to test for chloride and other common anions in water from a community purifier.

[94] 社区净水机
[95] 纯净水

扫一扫　视频 3-5　常见阴离子的分离和鉴定(讲解)
　　　　视频 3-6　常见阴离子的分离和鉴定(实验)

(徐孝菲编写)

Expt. 3　Determination of Reaction Rate Constant, Activation Energy and Influencing Factors
反应速率常数、活化能及影响因素的测定

Objectives

(1) Understand how factors such as concentration, temperature, and catalyst[96] affect the rate of chemical reactions.

(2) Master how to determine reaction rate[97] and activation energy[98].

(3) Use software such as Excel or Origin to process data and plot graphs.

Principles

Key terms: chemical kinetics[99], elementary reactions[100], law of mass action[101], reaction order[102], chemical reaction rate, chemical reaction rate constant, activation energy.

Chemical kinetics focuses on the rate and mechanisms of a reaction. In this experiment, we will study the kinetics of a redox reaction[103] between potassium persulfate ($K_2S_2O_8$) and potassium iodide (KI) and determine the reaction order and the activation energy. As shown in Reaction (3-1), $K_2S_2O_8$ is an oxidant[104], and KI is a reductant[105].

$$S_2O_8^{2-} + 2I^- = 2SO_4^{2-} + I_2 \qquad (3-1)$$

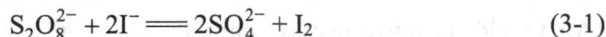

Since it is uncertain whether this reaction is an elementary reaction, we can begin by expressing the reaction rate as shown in Equation (3-2) and determine the reaction rate using experimental methods.

$$v = k[S_2O_8^{2-}]^m[I^-]^n \qquad (3-2)$$

In Equation (3-2), v is the instantaneous rate[106] of the reaction, measured in $mol \cdot L^{-1} \cdot s^{-1}$. k is the reaction rate constant, which is related to the properties of the reactants. k is independent of the concentration of the reactants but is strongly affected by temperature[107]. The unit of k is $(mol \cdot L^{-1})^{-(m+n-1)} \cdot s^{-1}$, and it depends on the reaction order. m and n are the reaction orders of the two reactants, respectively. The sum of m and n represents the overall reaction order[108].

[96] 催化剂
[97] 反应速率
[98] 活化能

[99] 化学动力学
[100] 基元反应
[101] 质量作用定律
[102] 反应级数
[103] 氧化还原反应

[104] 氧化剂
[105] 还原剂

[106] 瞬时速率
[107] k 为反应速率常数，与反应物本性有关，与反应物浓度无关，与温度有很大关系
[108] m 和 n 之和为总反应级数

From Equation (3-2), it can be concluded that the higher the concentration of the reactants, the greater the reaction rate will be at the same temperature[109].

At the beginning of the reaction, Equation (3-2) can be transformed into Equation (3-3), where $\overline{v}$ represents the average reaction rate, which can be measured using experimental methods.

$$\overline{v} = \frac{-\Delta[S_2O_8^{2-}]}{\Delta t} = k[S_2O_8^{2-}]^m[I^-]^n \tag{3-3}$$

Since the reactants and products in Reaction (3-1) are colorless, it is impossible to observe the progress of the reaction through visual observation. Therefore, we need to introduce another Reaction (3-4) as an indicator.

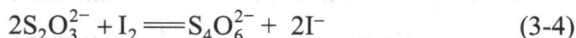

$$2S_2O_3^{2-} + I_2 = S_4O_6^{2-} + 2I^- \tag{3-4}$$

The rate of Reaction (3-4) is much faster than that of the main Reaction (3-1). When the amount of $Na_2S_2O_3$ is less than that of $K_2S_2O_8$, the progress of Reaction (3-1) can be indicated by starch. The I_2 produced by Reaction (3-1) quickly reacts with $Na_2S_2O_3$. After a period of time (Δt), the $Na_2S_2O_3$ is consumed, and the continuously generated I_2 from Reaction (3-1) immediately reacts with starch, turning the solution blue. When the solution turns blue, $Na_2S_2O_3$ has been completely consumed. At this point, $\Delta[S_2O_3^{2-}]$ is equal to its initial concentration[110].

Based on Reaction (3-4) and Reaction (3-1), we can re-write Equation (3-3) as Equation (3-5). Finally, Equation (3-6) is obtained by taking the logarithm to base 10[111] (lg) of both sides of Equation (3-5).

$$v = \overline{v} = -\frac{\Delta[S_2O_8^{2-}]}{\Delta t} = -\frac{\Delta[S_2O_3^{2-}]}{2\times\Delta t} = \frac{[S_2O_3^{2-}]_0}{2\times\Delta t} = k[S_2O_8^{2-}]^m[I^-]^n \tag{3-5}$$

$$\lg v = \lg\frac{\Delta[S_2O_3^{2-}]_0}{2\times\Delta t} = \lg k + m\lg[S_2O_8^{2-}] + n\lg[I^-] \tag{3-6}$$

The question is how to obtain the values of the reaction orders, m and n. When $[I^-]$ is kept constant, plotting $\lg v$ against $\lg[S_2O_8^{2-}]$ gives a linear graph, and the slope of the line equals m. Similarly, when $[S_2O_8^{2-}]$ is kept constant, plotting $\lg v$ against $\lg[I^-]$ gives a linear graph, and the slope of the line equals n. Finally, the reaction rate constant k can be obtained by substituting m and n into Equation (3-2)[112].

Temperature has a significant effect on the rate of chemical reactions. The reaction rate increases approximately 2 to 4 times for every 10ºC rise in temperature (van't Hoff's law)[113]. The quantitative relationship between the reaction rate constant and temperature can be expressed by Arrhenius' Equation (3-7).

[109] 相同温度下,反应物浓度越高,反应速率越大

[110] (实验设计思路)反应(3-1)为主反应且慢,反应(3-4)为辅助反应且快,加入 $Na_2S_2O_3$ 的量更少。因此,主反应(3-1)产生的 I_2 与 $Na_2S_2O_3$ 立即反应。当 I_2 和淀粉结合显示蓝色时,说明开始加入的 $Na_2S_2O_3$ 全部消耗完毕

[111] 以 10 为底数的对数

[112] (实验设计思路)保持 $[I^-]$ 不变,改变 $[S_2O_8^{2-}]$,测定反应速率,以 $\lg v$ 对 $\lg[S_2O_8^{2-}]$ 作图,直线斜率为 m。用同样方法可测得 n

[113] 温度每升高 10℃,反应速率提高 2~4 倍

$$k = A\mathrm{e}^{-\frac{E_\mathrm{a}}{RT}} \tag{3-7}$$

$$\ln k = -\frac{E_\mathrm{a}}{RT} + \ln A$$

or

$$\lg k = -\frac{E_\mathrm{a}}{2.303RT} + \lg A \tag{3-8}$$

In the above two equations, E_a is the activation energy of the reaction, with unit of $kJ\cdot mol^{-1}$. R is the molar gas constant[114], with the value of 8.314×10^{-3} $kJ\cdot mol^{-1}\cdot K^{-1}$. T is the thermodynamic temperature[115], measured in K. A is the pre-exponential factor[116], which is independent of temperature and concentration, and differs for various reactions.

By keeping $[I^-]$ and $[S_2O_8^{2-}]$ constant, the corresponding reaction rate constant (k) can be obtained by measuring the reaction time at different temperatures (T). Plot $\lg k$ versus $1/T$ according to Equation (3-8) to form a straight line, from which $\lg A$ can be determined from the intercept of the line, and the activation energy is calculated from the slope of the line[117].

Additionally, factors such as ionic strength and the presence of a catalyst[118] also influence the reaction rate.

Pre-lab Questions

(1) Briefly summarize the basic concepts of reaction rate, reaction rate constant, reaction order and activation energy.

(2) Briefly describe how to determine the reaction order and activation energy in this experiment.

Apparatus and Reagents

Apparatus: thermostatic waterbath[119], Erlenmeyer flask (150 mL), graduated cylinder, pipette, timer, thermometer.

Reagents: KNO_3 (0.20 $mol\cdot L^{-1}$), KI (0.20 $mol\cdot L^{-1}$), $K_2S_2O_8$ (0.10 $mol\cdot L^{-1}$), K_2SO_4 (0.10 $mol\cdot L^{-1}$), $Cu(NO_3)_2$ (0.02 $mol\cdot L^{-1}$), $Na_2S_2O_3$ (0.02 $mol\cdot L^{-1}$), starch (0.2%).

Procedures

To prevent contamination, any solution taken out should not be poured back into its original reagent bottle, and the dropper should not be inserted directly into the reagent bottle[120].

1. Effect of Concentration of K₂S₂O₈ on the Reaction Rate

Carry out this part of the experiment at room temperature, and measure

[114] 摩尔气体常量
[115] 热力学温度
[116] 指前因子

[117] (测定不同温度下的反应时间，通过作图得到活化能)
[118] 离子强度和催化剂等因素

[119] 恒温水浴锅

[120] 滴管不能直接伸入试剂瓶

and record the reaction time for each group in Table 3-6[121]. Required amounts of 0.10 $mol·L^{-1}$ $K_2S_2O_8$, 0.20 $mol·L^{-1}$ KI, 0.020 $mol·L^{-1}$ $Na_2S_2O_3$, 0.2% starch, 0.20 $mol·L^{-1}$ KNO_3 and 0.10 $mol·L^{-1}$ K_2SO_4 are listed in Table 3-6.

Table 3-6　Effect of concentration of $K_2S_2O_8$ on the reaction rate (T:＿＿℃)

No.	$V(K_2S_2O_8)$ /mL	$V(KI)$ /mL	$V(Na_2S_2O_3)$ /mL	$V(starch)$ /mL	$V(KNO_3)$ /mL	$V(K_2SO_4)$ /mL	V_{total} /mL	Δt /s
1	25	25	5.0	2.0	0.0	0.0	57	
2	20	25	5.0	2.0	0.0	5.0	57	
3	15	25	5.0	2.0	0.0	10	57	
4	10	25	5.0	2.0	0.0	15	57	
5	5.0	25	5.0	2.0	0.0	20	57	

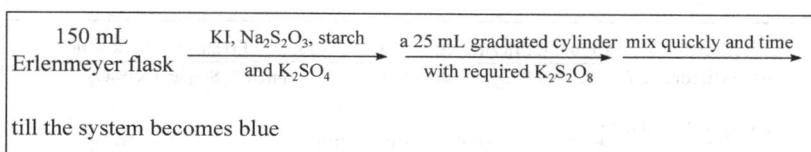

Note:

(1) To avoid contamination, use a separate graduated cylinder for $K_2S_2O_8$. Additionally, $K_2S_2O_8$ should not be added to the Erlenmeyer flask beforehand.

(2) The rate of shaking the Erlenmeyer flask affects the reaction rate, so try to maintain a consistent shaking rate each time.

2. Effect of Concentration of KI on the Reaction Rate

Complete the experiments in Table 3-7 following the above procedures at room temperature. Measure the reaction time for each group, and record the data in Table 3-7[122].

Table 3-7　Effect of concentration of KI on the reaction rate (T:＿＿℃)

No.	$V(K_2S_2O_8)$ /mL	$V(KI)$ /mL	$V(Na_2S_2O_3)$ /mL	$V(starch)$ /mL	$V(KNO_3)$ /mL	$V(K_2SO_4)$ /mL	V_{total} /mL	Δt /s
6	25	25	5.0	2.0	0.0	0.0	57	
7	25	20	5.0	2.0	5.0	0.0	57	
8	25	15	5.0	2.0	10	0.0	57	
9	25	10	5.0	2.0	15	0.0	57	
10	25	5.0	5.0	2.0	20	0.0	57	

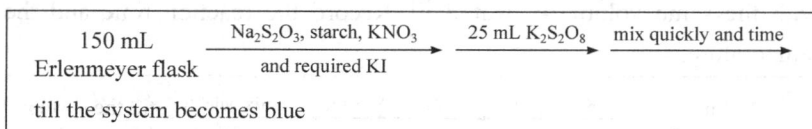

[121] (实验要点)用一个量筒专门量取 $K_2S_2O_8$ 溶液，且不能事先加入锥形瓶中。分别量取其他溶液并置于同一个锥形瓶中。准备完毕后，将 $K_2S_2O_8$ 溶液迅速倒入锥形瓶中，同时计时

[122] (实验要点同上)

3. Effect of Temperature on the Reaction Rate

[123] (无须一定按本书的
温度间隔，只要有准确的
升高温度即可)

Repeat this experiment in a thermostatic waterbath at approximately 5°C, 10°C, 15°C, and 20°C higher than room temperature, respectively[123].

Following the amount listed for No.3 in Table 3-6, add the required amounts of reagents into a 150 mL Erlenmeyer flask. Measure and record the reaction time and the temperature of the reaction system, then enter the data into Table 3-8.

Table 3-8   Effect of temperature on the reaction rate

No.	11	12	13	14
$T/°C$				
$\Delta t/s$				

water bath with different T $\xrightarrow{\text{a 150 mL Erlenmeyer flask with KI, Na}_2\text{S}_2\text{O}_3\text{, starch, KNO}_3}$ $\xrightarrow{\text{a 150 mL Erlenmeyer flask with required K}_2\text{SO}_4 \text{ and K}_2\text{S}_2\text{O}_8}$

$\xrightarrow{\text{mix quickly and time}}$ till the system becomes blue

Note:

(1) First, maintain the temperature constant for 5 minutes, then measure the temperature change before and after 1 minute[124]. If there is no change, the temperature is considered constant.

[124] 测定 1 min 前后的
温度

(2) Be careful not to maintain the constant temperature for too long, as the solvent may evaporate, leading to a change in the concentration of the reactants, which can affect the experimental results[125].

[125] 不能恒温太久，否
则溶剂蒸发将改变浓度，
影响反应速率

4. Effect of Catalyst on the Reaction Rate

At room temperature, follow the amounts and procedures for No.3 in Table 3-6, but first add 2 drops of $Cu(NO_3)_2$ as a catalyst to one of the Erlenmeyer flasks. Record the reaction time and the temperature.

150 mL Erlenmeyer flask $\xrightarrow{\text{KNO}_3\text{, KI, Na}_2\text{S}_2\text{O}_3\text{, K}_2\text{SO}_4\text{, starch, and 2 drops of Cu(NO}_3)_2}$ $\xrightarrow{\text{25 mL K}_2\text{S}_2\text{O}_8}$ $\xrightarrow{\text{mix quickly and time}}$ till the system becomes blue

5. Effect of Ionic Strength on the Reaction Rate

At room temperature, follow the amounts and procedures for No.3 in Table 3-6 and No.8 in Table 3-7, but replace the K_2SO_4 or KNO_3 solution with the same volume of water[126]. Record the reaction time and the temperature.

[126] 用相同体积的水代
替两种溶液

150 mL Erlenmeyer flask $\xrightarrow{\text{KI, Na}_2\text{S}_2\text{O}_3 \text{ starch and water}}$ $\xrightarrow{\text{25 mL K}_2\text{S}_2\text{O}_8}$ $\xrightarrow{\text{mix quickly and time}}$ till the system becomes blue

Data Treatment and Analysis

(1) Calculate the initial $K_2S_2O_8$ concentration ($[K_2S_2O_8]_0$), the reaction rate, and the reaction rate constant for each experiment in Table 3-6. Then use computer software to plot $\lg v$ versus $\lg[K_2S_2O_8]_0$, and obtain the linear equation and R^2 of the best fit line[127]. From this, m can be determined. Record the results in Table 3-9.

[127] 最佳拟合直线

Table 3-9   Effect of concentration of $K_2S_2O_8$ on the reaction rate (T:_____°C)

No.	1	2	3	4	5
$[K_2S_2O_8]_0/(\text{mol·L}^{-1})$					
$\Delta t/\text{s}$					
$v/(\text{mol·L}^{-1}\cdot\text{s}^{-1})$					
$\lg[K_2S_2O_8]_0$					
$\lg v$					

(2) Calculate the initial KI concentration ($[KI]_0$), the reaction rate, and the reaction rate constant for each experiment in Table 3-7. Then use computer software to plot $\lg v$ versus $\lg[KI]_0$, and obtain the linear equation and R^2 of the best fit line. From this, n can be determined. Record the results in Table 3-10.

Table 3-10   Effect of concentration of KI on the reaction rate (T:_____°C)

No.	6	7	8	9	10
$[KI]_0/(\text{mol·L}^{-1})$					
$\Delta t/\text{s}$					
$v/(\text{mol·L}^{-1}\cdot\text{s}^{-1})$					
$\lg[KI]_0$					
$\lg v$					

(3) Calculate the data in Table 3-11 based on the original data from Table 3-8. Then use computer software to plot $\ln k$ versus $1/T$ and obtain the linear equation and R^2 of the best fit line. From this, E_a can be calculated[128].

[128] 作 $\ln k$-$1/T$ 的线性拟合曲线，从而获得活化能

Table 3-11   Effect of temperature on the reaction rate

No.	3	11	12	13	14
$T/°C$					
$\Delta t/\text{s}$					
$v/(\text{mol·L}^{-1}\cdot\text{s}^{-1})$					
$1/T/\text{K}^{-1}$					
k					
$\ln k$					

(4) Summarize and explain the effects of the catalyst and ionic strength on the reaction rate, respectively.

Safety and Waste Disposal

(1) Wear a lab coat and safety goggles at all times in the laboratory. Wear gloves when handling hot glassware. If any chemicals splash into eyes or onto skin, wash with water immediately.

(2) The iodine (I_2) produced in the experiment has potential toxicity. Iodine on skin can be washed off with water or a $Na_2S_2O_3$ or Na_2CO_3 solution.

(3) $K_2S_2O_8$ is a strong oxidant, so be careful when using it.

(4) Never pour the oxidant or reductant waste solutions down the sink, react the solutions with each other first.

Post-lab Questions

(1) Taking this experiment as an example, can the reaction order be directly obtained from the chemical reaction equation?

(2) Why can the reaction rate be calculated by the time the reaction solution appears blue?

Exploring Experiments

Design an experiment to determine the activation energy when using a catalyst. Compare the value without the catalyst and discuss its catalytic effect on the reaction.

扫一扫　视频 3-7　速率常数和活化能的测定(讲解)
　　　　视频 3-8　速率常数和活化能的测定(实验)
　　　　视频 3-9　用计算机处理数据和作图

(何桂金编写)

Expt. 4　Preparation of Alum and Large Crystal Cultivation
明矾的制备及其大晶体培养

Objectives

[129] 复盐

(1) Understand the concept and characteristics of double salts[129].

(2) Learn the mechanism and process of alum synthesis, using Al powder.

(3) Practice basic operations, such as dissolution, crystallization, and filtration[130].

[130] 溶解、结晶和过滤

Principles

Key terms: amphoteric compound[131], double salts, solubility curve[132], crystal growth[133].

[131] 两性物质
[132] 溶解度曲线
[133] 晶体生长

Aluminum potassium sulfate, $KAl(SO_4)_2·12H_2O$ (also known as alum[134]), is prepared by mixing　aluminum sulfate $[Al_2(SO_4)_3]$ with potassium sulfate (K_2SO_4) in water solution. Alum is a colorless crystal that easily dissolves in water. It loses 9 crystallization water molecules[135] at 64.5℃ and 12 crystallization water molecules at 200℃. The hydration[136] of alum can produce colloidal[137] aluminum hydroxide $[Al(OH)_3]$, which can adsorb suspended particles in water to form deposits. Therefore, alum is often used for water purification.

[134] 明矾
[135] 结晶水分子
[136] 水解
[137] 胶状的

In this experiment, aluminum (Al) powder will be dissolved in a sodium hydroxide (NaOH) solution to obtain a sodium tetrahydroxy-aluminate $[NaAl(OH)_4]$ solution. This chemical reaction can be written as Equation (3-9).

$$2Al + 2NaOH + 6H_2O = 2NaAl(OH)_4 + 3H_2\uparrow \qquad (3\text{-}9)$$

Some impurities in the aluminum powder are insoluble in NaOH solution and can be removed by filtration[138]. After removing impurities, the pH of the $NaAl(OH)_4$ solution is adjusted to 8-9 with sulphuric acid (H_2SO_4) to form an aluminum hydroxide $[Al(OH)_3]$ precipitate. The collected $Al(OH)_3$ precipitate is further treated with more H_2SO_4 to obtain an $Al_2(SO_4)_3$ solution. Finally, by adding potassium sulfate (K_2SO_4), alum crystals precipitate due to the low solubility in water when cooled to room temperature (Table 3-12).

[138] 过滤除去

$$2NaAl(OH)_4 + H_2SO_4 = 2Al(OH)_3\downarrow + Na_2SO_4 + 2H_2O \quad (3\text{-}10)$$

$$3H_2SO_4 + 2Al(OH)_3 = Al_2(SO_4)_3 + 6H_2O \qquad (3\text{-}11)$$

$$Al_2(SO_4)_3 + K_2SO_4 + 24H_2O = 2KAl(SO_4)_2·12H_2O\downarrow \quad (3\text{-}12)$$

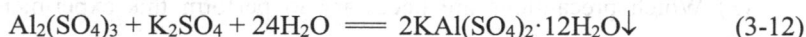

Table 3-12　Solubility of related salts at different temperatures [unit: g·(100 g water)$^{-1}$]

T/K	273	283	293	303	313	323	333	343	353	363	373
alum	3.00	3.99	5.90	8.39	11.7	17	24.8	40	71.0	109	154
K_2SO_4	7.4	9.3	11.1	13.0	14.8	16.5	18.2	19.8	21.4	22.9	24.1
$Al_2(SO_4)_3$	31.2	33.5	36.4	40.4	45.7	52.2	59.2	66.2	73.1	86.8	89

Fig. 3-6 shows the typical solubility curves of an inorganic salt[139].

[139] 典型无机盐的溶解度曲线

The BB_1 curve is a normal solubility curve, and the CC_1 curve is a supersaturated (oversaturated) curve[140]. The region under BB_1 represents the unsaturated zone[141]. To obtain crystals when an unsaturated solution is at point A, two methods can be used: the $A{\rightarrow}B$ cooling method or the $A{\rightarrow}B_1$ concentration method[142].

[140] 过饱和曲线
[141] 不饱和区域

[142] 浓缩法

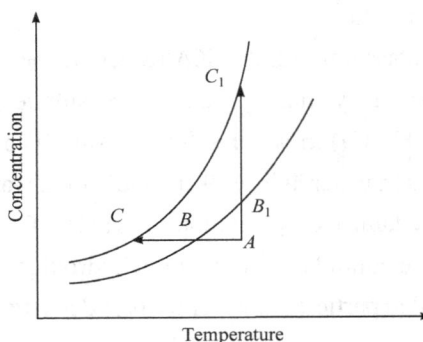

Fig. 3-6　Solubility curves

These methods can change the state of the solution to the zone above the BB_1 curve. Sometimes the solution enters an oversaturated state, but no crystal precipitation occurs. However, the oversaturated state is highly unstable. In this state, new crystals can form with slight motions (such as rubbing the internal surface of the container[143]). The region between the CC_1 and BB_1 curve is defined as the metastable zone[144]. To obtain a regular, large crystal, the solution should be kept in the metastable state, and the crystal seed[145] growth should be controlled at a slow rate to avoid the precipitation of small crystals.

[143] 摩擦容器内壁
[144] 介稳区

[145] 晶种

Pre-lab Questions

(1) Explain the concepts of amphoteric compounds and their properties.

(2) Explain the concepts of double salts and their solubility properties.

(3) Which precautions are necessary to perform this experiment safely?

Apparatus and Reagents

Apparatus: hot plate, vacuum pump[146], suction flask[147], Buchner funnel, thermometer[148].

[146] 真空泵
[147] 抽滤瓶
[148] 温度计

Reagents: Al powder (s), K_2SO_4 (s), NaOH (s, 2 mol·L^{-1}), H_2SO_4 (1 : 1, 3 mol·L^{-1}), alum seed crystals (prepare an alum solution 20-30℃ above room temperature, then pour the saturated solution into a large flat-bottomed enamel tray until the liquid level is 2-3 cm high.

After cooling for 24 hours and allowing plenty of small crystals to precipitate at the bottom, select the perfect crystals as crystals seed[149]).

[149] 选择完美的晶体作为晶种

Procedures

1. Preparation of Al(OH)₃

Ⅰ. Preparation of NaAl(OH)₄

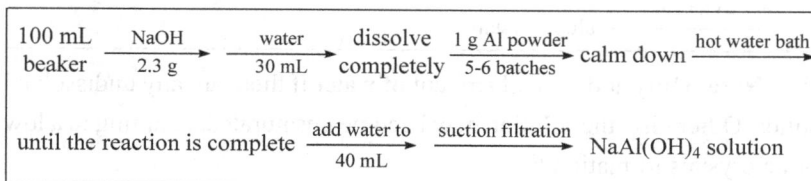

Note:

(1) Large amounts of heat and bubbles will rise. Therefore, it is necessary to add the Al powder in 5-6 portions while constantly stirring; otherwise, the solution may boil over[150].

(2) Heat the beaker in a hot water bath only after the reaction system has calmed down[151].

[150] 产生大量的热和气泡。因此，铝粉要分批加入且不断搅拌，否则易暴沸

[151] 反应体系平稳后方可水浴加热

Ⅱ. Formation of Al(OH)₃

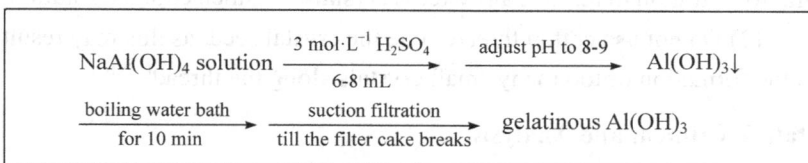

Note:

(1) If pH< 8, adjust it with 2 mol·L⁻¹ NaOH solution.

(2) pH adjustment is crucial; otherwise, the precipitation will be incomplete, decreasing the yield[152].

(3) Al(OH)₃ colloidal particles are small and difficult to filter. Aging can effectively increase the particle size, so avoid stiring during the aging process[153].

(4) When washing the filter cake, do not stir it to prevent damaging the filter paper. Continue suction filtration until the filter cake breaks[154].

[152] pH 调节是关键，否则沉淀不完全，产率降低

[153] Al(OH)₃ 胶状颗粒小，难以过滤。陈化可以增大颗粒尺寸，但搅拌会打断颗粒长大的过程，故不能搅拌

[154] 抽滤至滤饼裂开

2. Preparation of Al₂(SO₄)₃

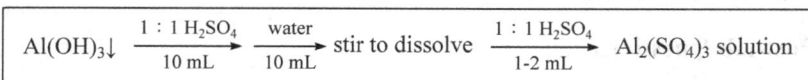

Note:

(1) If Al(OH)₃ does not dissolved completely, add an addition 1-

2 mL H_2SO_4 (1：1). However, too much H_2SO_4 will make it more difficult to crystalize the alum[155].

(2) If the solution remains cloudy, filtrate the solution and use the filtrate for the following steps[156].

3. Preparation of Alum

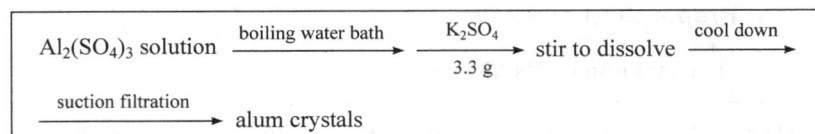

$Al_2(SO_4)_3$ solution $\xrightarrow{\text{boiling water bath}}$ $\xrightarrow[\text{3.3 g}]{K_2SO_4}$ stir to dissolve $\xrightarrow{\text{cool down}}$ $\xrightarrow{\text{suction filtration}}$ alum crystals

Note: Only add a small amount of water if there are any undissolved solids. Otherwise, the solution may become unsaturated, resulting in a low or no crystals formation[157].

4. Cultivation of Big Alum Crystals

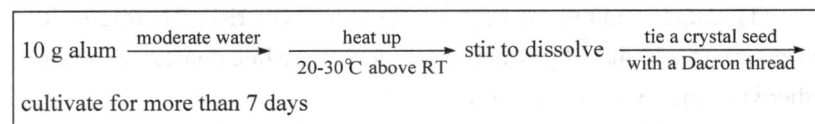

10 g alum $\xrightarrow{\text{moderate water}}$ $\xrightarrow[\text{20-30℃ above RT}]{\text{heat up}}$ stir to dissolve $\xrightarrow[\text{with a Dacron thread}]{\text{tie a crystal seed}}$ cultivate for more than 7 days

Note:

(1) Gently heat the alum crystals until dissolved. Afterward, use gravity filtration to remove any excess crystals[158], then cool the solution.

(2) Do not use cotton thread to tie the crystal seed, as this may result in the formation of too many small crystals along the thread[159].

Data Treatment and Analysis

Calculate both the theoretical mass and percent yield of the product.

Safety and Waste Disposal

(1) Wear a lab coat and safety goggles at all times in the laboratory. If any chemicals splash into eyes or onto skin, wash with water immediately.

(2) NaOH, $Al(OH)_3$ and H_2SO_4 are corrosive, they must be handled with care. Wear gloves whenever needed.

(3) Concentrated aluminum (Al) is very toxic. It can deposit in the brain, causing memory loss, mental decline, and even Alzheimer's disease[160].

(4) Never pour the acidic or alkaline waste solution down the sink, neutralize the solutions first.

Post-lab Questions

(1) Why should the solution be adjusted to a pH of 8-9 to form $Al(OH)_3$?

[155] 过多硫酸会使明矾晶体难析出
[156] 若溶液仍然浑浊，过滤后取滤液用于下列步骤
[157] 如果有不溶固体，只能补加一点点水。否则可能成为不饱和溶液，导致很少或没有晶体生成
[158] 常压过滤除去多余的晶体
[159] 不能用棉线绑晶种，否则棉线上会有很多小晶体生成(棉线上的纤维可为晶核中心)
[160] 高浓度铝具有毒性，会在大脑中沉积，导致记忆丧失、智力下降，甚至引起阿尔茨海默病(老年痴呆症)

(2) What is the best way to get a large alum crystal with perfect shape and good transparency?

Exploring Experiments

1. Determining Al Content in Alum by Back Titration[161]

(1) Weigh 0.20-0.25 g of alum sample (M_r = 474.4, to the nearest 0.0001 g) and place it into a 100 mL beaker. Add 40 mL of water and stir until the solids completely dissolve. Then, transfer the solution into a 100 mL volumetric flask and dilute to the mark.

(2) Pipette 20.00 mL of the alum solution into an Erlenmeyer flask, add 20.00 mL of 0.01 mol·L^{-1} standard EDTA solution, and 15 mL of HAc-NaAc buffer solution[162] (pH 4.3). Boil the solution for 1-2 minutes, then add 2-5 drops of PAN indicator.

(3) Titrate with a standard CuSO$_4$ solution until the solution changes from yellow to bright purple.

2. Determining Al Content in Alum by Direct Titration

(1) Weigh 0.10-0.15 g of alum sample (M_r = 474.4, to the nearest 0.0001g) and place it into an Erlenmeyer flask. Add 25 mL of water and swirl the flask until the solids are completely dissolved. Add 20.00 mL of standard 0.02 mol·L^{-1} EDTA solution and 2 drops of xylenol orange indicator[163]. Carefully add NH$_3$·H$_2$O (1∶1) until the solution turns purple. Then, add 3 drops of HCl (1∶3) and boil the solution for 3 min.

(2) After the solution cools down, add 20 mL of 20% hexamethylene tetramine[164]. The solution should appear yellow or orange. If not, add HCl (1∶3) dropwise until the solution reaches the desired color. Add another 2 drops of xylenol orange, and titrate with standard Zn^{2+} solution until the mixture turns purplish red (the titrated volume is not counted). Then add 1 mL of 20% NH$_4$F and shake well.

(3) Heat the solution to a gently boil, and after cooling, add 2 drops of xylenol orange after the solution cools down. The solution should be yellow or orange. If not, add HCl (1∶3) dropwise until the solution reaches the desired color.

(4) Titrate with a standard Zn^{2+} solution until the mixture turns purplish red.

[161] 返滴定法

[162] 缓冲溶液

[163] 二甲酚橙指示剂

[164] 六次甲基四胺

扫一扫　视频 3-10　明矾的制备及其大晶体培养(讲解)
视频 3-11　明矾的制备及其大晶体培养(实验)

(何桂金编写)

Chapter 4   Fundamental Analytical Chemistry Experiments
基础分析化学实验

Expt. 5 Preparation and Standardization of the Solutions for Acid-base Titration
酸碱滴定溶液的配制与标定

Introduction to Acid-base Titration

[1] 转化成酸或碱

Acid-base titration is a type of titration based on acid-base reactions. It can be applied to determine substances that can undergo acid-base reactions or convert into an acid or a base[1]. The acid-base titration curve is a plot of pH versus the volume of titrant added (pH-$V_{titrant}$). A titration jump shows the rapid pH change during the final stage of the reaction, from 99.9% to 100.1% of the stoichiometric point[2].

[2] 酸碱滴定曲线是以 pH 对滴定剂体积作图。滴定突跃表明化学计量点前后±0.1%的急剧 pH 变化

$$NaOH + HCl \rlap{=}= NaOH + H_2O \text{ (a strong base with a strong acid)}$$
$$NaOH + HAc \rlap{=}= NaAc + H_2O \text{ (a strong base with a weak acid)}$$
$$NaOH + H_3PO_4 \rlap{=}= NaH_2PO_4 + H_2O \text{ (a strong base with a polyprotic acid}[3])$$
$$2NaOH + H_2C_2O_4 \rlap{=}= Na_2C_2O_4 + H_2O \text{ (a strong base with a polyprotic acid)}$$

[3] 强碱和多元酸

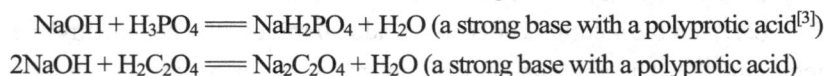

Influences on the Titration Jump

(1) The titration jump of a strong acid with a strong base depends on their concentrations. The higher the concentrations, the larger the jump (Fig. 4-1).

[4] 弱酸与强碱的滴定突跃和它们的浓度及解离常数有关

(2) The titration jump of a weak acid with a strong base depends on both the concentrations and the dissociation constant[4]. The higher the concentrations and dissociation constant, the larger the jump (Fig. 4-2).

(3) The titration jump of a weak polyprotic acid with a strong base is more complicated.

[5] 直接准确滴定和分步滴定的判定依据

Some Criteria for Effective and Stepwise Titrations[5]

[6] 对于准确滴定分析，滴定误差不大于±0.2%

(1) Criteria for effective titrations: For accurate titrimetric analysis, the titration error should be less than $\pm 0.2\%$[6]. To achieve this, the cK value of the titrated acid or base should be greater than 10^{-8} (i.e. $cK \geqslant 10^{-8}$).

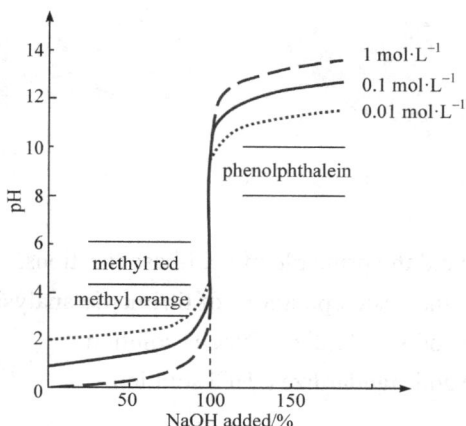

Fig. 4-1   A strong acid with a strong base

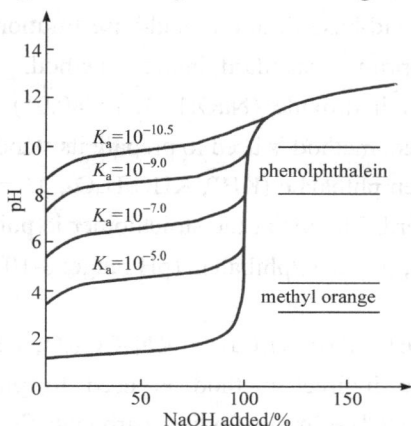

Fig. 4-2   Some weak acids with a strong base

(2) Criteria for stepwise titrations: If the required titration error is less than $\pm 0.2\%$, protons in polyprotic acids can be titrated stepwise when $K_n/K_{n+1} \geqslant 10^4$.

Acid-base Titration Indicator

(1) Structure of an indicator: These indicators are weak organic acids or bases (organic dyes), with the acidic or alkaline forms exhibiting different colors from their conjugate pairs[7].

(2) Color change of an indicator: The color of an indicator depends on $c(H^+)$ (pH)[8]. Theorically, the point of pH change occurs at the indicator's pK_{HIn}, and the effective pH range is $pK_{HIn} \pm 1$.

(3) Selection of an indicator: An appropriate indicator should change color at the same pH as the stoichiometric point of the acid-base reaction. The pH range of the indicators must be narrower than, or within the titration jump[9]. For easier detection with the naked eyes, the indicator should show a color change from light to dark or from colorless to colored.

[7] 有机弱酸或弱碱，其颜色与它们的共轭对不同

[8] 指示剂颜色与 H^+浓度 (pH)有关

[9] 指示剂的pH范围应部分或全部落在滴定突跃内

扫一扫　视频 4-1　酸碱滴定法概述

Objectives

(1) Understand the principle of acid-base titrations.

(2) Practice the basic operations of titrimetric analysis.

(3) Prepare and standardize a NaOH solution.

(4) Prepare and standardize a HCl solution.

Principles

Key terms: acid-base titration, acid-base titration curve, acid-base titration indicator, primary standard, indirect method.

Since sodium hydroxide (NaOH, $M_r = 40.01$) is not a primary standard, the indirect method is used to prepare its standard solution, with potassium hydrogen phthalate (KHP, $KHC_8H_4O_4$, $M_r = 204.2$) serving as the primary standard. The pH at the stoichiometric point of this reaction is approximately 9, so phenolphthalein (pH range: 8-10) can be used as an indicator[10].

$$NaOH + KHC_8H_4O_4 = KNaC_8H_4O_4 + H_2O$$

Similarly, the indirect method is used to prepare a standard hydrochloric acid (HCl) solution. Sodium carbonate (Na_2CO_3, $M_r = 106.0$) or borax[11] ($Na_2B_4O_7 \cdot 10H_2O$, $M_r = 381.4$) can be used as primary standards. Methyl orange[12] (pH range: 3.1-4.4) or methyl red[13] (pH range: 4.4-6.2) can be used as the indicator.

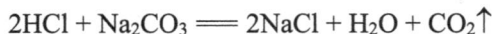

$$2HCl + Na_2CO_3 = 2NaCl + H_2O + CO_2\uparrow$$
$$Na_2B_4O_7 + 2HCl + 5H_2O = 4H_3BO_3 + 2NaCl$$

Pre-lab Questions

(1) What are the definition and characteristics of a primary standard?

(2) Briefly describe the different ways to prepare 100 mL of a standard $0.1 \ mol \cdot L^{-1}$ solution, taking potassium hydrogen phthalate (KHP) and NaOH for examples.

Apparatus and Reagents

Apparatus: analytical balance, hot plate, burette (25 mL), volumetric flask (100 mL), pipette (20 mL), Erlenmeyer flask (3×150 mL), graduated cylinder, beaker, reagent bottle (2×250 mL, one with rubber stopper, another with glass stopper[14]).

[10] 间接法配制。标定时以邻苯二甲酸氢钾为基准物质，酚酞为指示剂

[11] 硼砂

[12] 甲基橙

[13] 甲基红

[14] 一个带橡胶塞，另一个带玻璃塞

Reagents: potassium hydrogen phthalate (s, primary standard), Na_2CO_3 (s, primary standard), $Na_2B_4O_7 \cdot 10H_2O$ (s, primary standard), NaOH (s), HCl (concentrated, 6 mol·L^{-1}), phenolphthalein (0.1%), methyl orange (0.1%), methyl red (0.1%).

Procedures

1. Preparation and Standardization of 0.1 mol·L^{-1} NaOH Solution

Step 1. Preparation of 250 mL 0.1 mol·L^{-1} NaOH solution.

100 mL beaker $\xrightarrow[1.0\ g]{NaOH}$ $\xrightarrow[50\ mL]{H_2O}$	250 mL reagent bottle	$\xrightarrow[200\ mL]{H_2O}$	stopper and shake

Note: Use a clean 100 mL beaker to weigh NaOH pellets, as they are highly corrosive and prone to deliquescence[15].

[15] 用烧杯称量 NaOH 固体，因为其腐蚀性强且易潮解

Step 2. Standardization of 0.1 mol·L^{-1} NaOH solution.

150 mL Erlenmeyer flask $\xrightarrow[0.30\text{-}0.40\ g]{KHC_8H_4O_4}$ $\xrightarrow[50\ mL]{H_2O}$ swirl until solids dissolve completely $\xrightarrow[1\text{-}2\ drops]{phenolphthalein}$

titrate with NaOH $\longrightarrow$ from colorless to pink (no change within 30 s) $\longrightarrow$ read the final volume

$\longrightarrow$ repeat two more times

Note:

(1) Do not swirl the Erlenmeyer flask violently because the NaOH solution tends to absorb atmospheric carbon dioxide[16].

[16] 不能剧烈振摇锥形瓶，因为 NaOH 容易吸收空气中二氧化碳

(2) Do not swirl the Erlenmeyer flask once the endpoint reached. The solution should turn colorless quickly.

2. Preparation and Standardization of 0.1 mol·L^{-1} HCl Solution

Step 1. Preparation of 250 mL 0.1 mol·L^{-1} HCl solution.

Method 1: Preparing by diluting concentrated HCl solution[17].

[17] 用浓 HCl 稀释

250 mL reagent bottle $\xrightarrow[248\ mL]{H_2O}$ $\xrightarrow[2.4\ mL]{conc.\ HCl}$ stopper and shake

Note: Because concentrated (conc.) HCl solution is highly corrosive and prone to spill[18], it is advisable to avoid using concentrated HCl solution for preparing this solution.

[18] 腐蚀性强且易溅出

Method 2: Preparing by diluting HCl stock solution.

250 mL reagent bottle $\xrightarrow[245\ mL]{H_2O}$ $\xrightarrow[4.5\ mL]{6\ mol·L^{-1}\ HCl}$ stopper and shake

Step 2. Standardization of 0.1 mol·L⁻¹ HCl solution.

Method 1: Using Na₂CO₃ as a primary standard.

150 mL Erlenmeyer flask	$\xrightarrow[\text{0.10-0.12 g}]{\text{Na}_2\text{CO}_3}$	$\xrightarrow[\text{20-30 mL}]{\text{H}_2\text{O}}$	swirl until solids dissolve completely	$\xrightarrow[\text{1-2 drops}]{\text{methyl orange}}$
titrate with HCl	$\longrightarrow$	from yellow to orange (no change within 30 s)	$\longrightarrow$	read the final volume
$\longrightarrow$	repeat two more times			

Note:

(1) Shake the Erlenmeyer flask vigorously to remove CO₂ from the solution; otherwise, the endpoint may be delayed[19].

(2) Use a "0.05 mol·L⁻¹ NaHCO₃ + indicator" solution as a reference solution to help to identify the endpoint[20].

(3) For better identification of the endpoint, a mixed indicator of methyl orange-bromocresol green[21] can be used. It changes from green to red at the endpoint.

[19] 否则，滴定终点将推后

[20] 作为参比溶液来帮助判断终点

[21] 甲基橙-溴甲酚绿混合指示剂

Method 2: Using Na₂B₄O₇·10H₂O as a primary standard.

150 mL Erlenmeyer flask	$\xrightarrow[\text{0.28-0.38 g}]{\text{Na}_2\text{B}_4\text{O}_7\cdot10\text{H}_2\text{O}}$	$\xrightarrow[\text{20-30 mL}]{\text{H}_2\text{O}}$	swirl until solids dissolve completely	$\xrightarrow[\text{1-2 drops}]{\text{methyl red}}$
titrate with HCl	$\longrightarrow$	from yellow to orange (no change within 30 s)	$\longrightarrow$	read the final volume
$\longrightarrow$	repeat two more times			

Data Treatment and Analysis

Calculate the accurate concentration of the NaOH and HCl solution. Calculate and analyze the relative average deviation[22].

[22] 计算并分析相对平均偏差

Safety and Waste Disposal

(1) Wear a lab coat and safety goggles at all times in the laboratory. Wear gloves when handling the hot glassware. If any chemicals splash into eyes or onto skin, wash with water immediately.

(2) NaOH solid and concentrated HCl are highly corrosive, they must be handled with care. Operate with the concentrated HCl in the fume hood. Wear gloves whenever needed.

(3) Do not pour the remaining acidic or alkaline solution down the sink. First neutralize them with each other.

Post-lab Questions

If the following circumstances occurred during the titration process, analyze their effect on the concentration of NaOH solution using KHP as the primary standard.

(1) Bubbles came out from the tip of the burette.

(2) Swirled the volumetric flask violently.

(3) Read the final volume of the burette from the upper side of the meniscus[23].

(4) The primary standard was a little wet.

(5) Forgot to rinse the burette with NaOH solution.

[23] 从弯月面上方读取滴定管的终体积

扫一扫　视频 4-2　NaOH 溶液的配制与标定
　　　　视频 4-3　HCl 溶液的配制与标定

(曾秀琼编写)

Expt. 6　Preparation and Standardization of the Solutions for Redox Titration
氧化还原滴定溶液的配制与标定

Introduction to Redox Titration

Redox titration is a type of titration based on oxidation-reduction reactions. It is primarily suitable for oxidizing or reducing agents, and secondarily, for substances that react with oxidizing or reducing agents.

The redox titration curve is a plot of E (potential of oxidant)-V (volume of titrant)[24]. A titration jump represents the potential change from 99.9% to 100.1% completion (Fig. 4-3). The magnitude of the titration jump depends on the potential difference between the oxidizing and reducing agents[25]. The larger the difference, the larger the jump.

[24] 以 E (氧化剂电势)对 V (滴定剂体积)作图

[25] 滴定突跃大小取决于氧化剂和还原剂之间的电势差

$$Ce^{4+} + Fe^{2+} =\!=\!= Ce^{3+} + Fe^{3+}$$

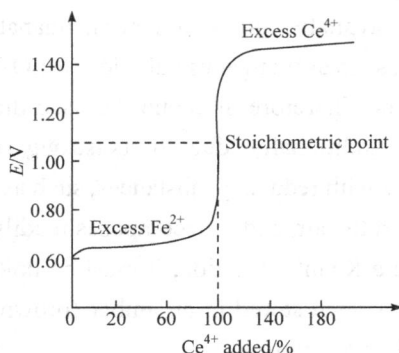

Fig. 4-3　Redox titration curve

Reduction half-reactions of this reaction are shown as following.

$$Ce^{4+} + e \longrightarrow Ce^{3+} \qquad E_1^{\ominus} = 1.44 \text{ V}$$
$$Fe^{3+} + e \longrightarrow Fe^{2+} \qquad E_2^{\ominus} = 0.68 \text{ V}$$

Redox Indicators

[26] 还原态和氧化态的颜色各异

In general, redox indicators are organic oxidants or reductants. The reduced state has one color, while the oxidized state has another[26]. There are three types of redox indicators.

[27] 自身指示剂

(1) Self indicators[27], such as a KMnO₄ solution, which changes color from purplish red to pink (light red) during the titration.

[28] 特殊指示剂，如淀粉溶液与碘反应生成深蓝色配合物

(2) Special indicators, like starch solution, which reacts with iodine (I_2) to form a dark blue complex[28].

(3) Indicators that participate in redox reaction.

扫一扫　视频 4-4　氧化还原滴定法概述

Part 1. Preparation and Standardization of 0.02 mol·L⁻¹ KMnO₄ Solution

Objectives

(1) Understand the principle of redox titrations.

[29] 高锰酸钾滴定法

(2) Understand the principle of permanganometric titration[29].

(3) Prepare and standardize a KMnO₄ solution.

Principles

Key terms: redox titration, permanganometric titration (permanganometry), self indicator, standard KMnO₄ solution.

[30] 氯化物、硫酸盐、硝酸盐

Commercially available potassium permanganate (KMnO₄) often contains impurities, such as manganese dioxide (MnO_2), chloride, sulfate, nitrate[30], and others. Therefore, it cannot be used directly to prepare a standard solution. Additionally, due to its strong oxidization ability, KMnO₄ easily reacts with reducing substances, such as organic impurities in water and ashes in the air, and it decomposes readily when exposed to

[31] 见光易分解

light[31]. To prepare a KMnO₄ solution, it must be boiled or dissolved in cold deionized water, then stored in an amber (brown) reagent bottle in the dark for several days before use.

[32] 滴定剂

Permanganometry: This is one of the redox titration methods, typically conducted in strongly acidic condition. A KMnO₄ solution is used as both the titrant[32] and the oxidizing agent, The MnO_4^- ions serve as a self indicator, eliminating the need for an additional indicator. A slightly excessive of $KMnO_4$ solution after the endpoint will turn the solution pink[33].

$$2MnO_4^- + 5C_2O_4^{2-} + 16H^+ === 2Mn^{2+} + 10CO_2\uparrow + 8H_2O$$

The following are three important conditions for standardizing $KMnO_4$ solution with the primary standard, sodium oxalate ($Na_2C_2O_4$)[34]:

(1) Acidity: The acidity of the solution should be 0.5-1.0 $mol \cdot L^{-1}$, as $KMnO_4$ is a strong oxidant in acidic solution.

(2) Temperature: The solution temperature should be 75-85℃. This is a self-catalyzed reaction that starts slowly. Higher temperatures accelerate the reaction.

(3) Titration speed: The titration should begin slowly, then proceed fast, and finally slow down again. At the start, $KMnO_4$ solution must be added dropwise due to the slow reaction. As Mn^{2+} is produced during titration, the reaction rate increases, allowing for a faster addition of $KMnO_4$. However, avoid adding $KMnO_4$ solution in a continuous stream, as it may decompose or react with Mn^{2+} rather than with $C_2O_4^{2-}$ in the hot acidic solution. Near the endpoint, the $KMnO_4$ solution should again be added dropwise.

Pre-lab Questions

(1) How do you prepare a standard $KMnO_4$ solution? Provide a brief description.

(2) For the standardization of a $KMnO_4$ solution, what are the effects when the following conditions are higher or lower: temperature, acidity, and titration speed? Give a brief explanation.

Apparatus and Reagents

Apparatus: analytical balance, hot plate, brown burette (25 mL), Erlenmeyer flask (3×150 mL), graduated cylinder, beaker, brown reagent bottle (1 L).

Reagents: $Na_2C_2O_4$ (s, primary standard, $M_r = 134.0$), $KMnO_4$ (s, $M_r = 158.0$), H_2SO_4 (3 $mol \cdot L^{-1}$).

Procedures

1. Preparation of 0.02 $mol \cdot L^{-1}$ $KMnO_4$ Solution

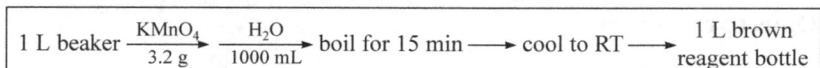

$$\boxed{\text{1 L beaker} \xrightarrow[\text{3.2 g}]{KMnO_4} \xrightarrow{\substack{H_2O \\ 1000\ mL}} \text{boil for 15 min} \longrightarrow \text{cool to RT} \xrightarrow{} \substack{\text{1 L brown} \\ \text{reagent bottle}}}$$

Note: $KMnO_4$ solution should be stored in a brown reagent bottle[35].

[33] 无须额外指示剂。终点后稍过量的 $KMnO_4$ 使溶液变粉色

[34] 用草酸钠标定 $KMnO_4$ 的三个条件：(1)酸度。(2)温度：因为自催化反应，Mn^{2+} 为催化剂。(3)滴定速度：先慢(逐滴加入)后快(但不呈流水状)再慢(逐滴加入)

[35] 棕色试剂瓶

2. Standardization of 0.02 mol·L^{-1} KMnO$_4$ Solution

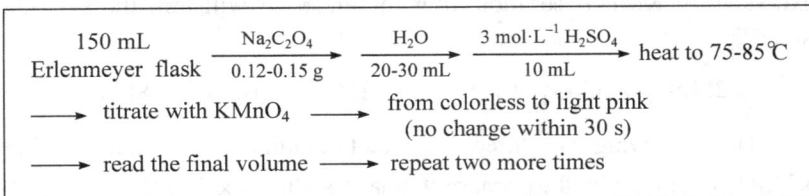

150 mL Erlenmeyer flask $\xrightarrow[\text{0.12-0.15 g}]{\text{Na}_2\text{C}_2\text{O}_4}$ $\xrightarrow[\text{20-30 mL}]{\text{H}_2\text{O}}$ $\xrightarrow[\text{10 mL}]{\text{3 mol·L}^{-1}\text{ H}_2\text{SO}_4}$ heat to 75-85 ℃

$\longrightarrow$ titrate with KMnO$_4$ $\longrightarrow$ from colorless to light pink (no change within 30 s)

$\longrightarrow$ read the final volume $\longrightarrow$ repeat two more times

Note:

(1) Use a brown burette to hold the KMnO$_4$ solution, and carefully follow the three conditions during the titration.

(2) Do not use tap water to wash the burette after titration, as it may cause the burette to turn brown. Instead, wash it with deionized water[36].

[36] 此处不能用自来水清洗滴定管，需用去离子水

Data Treatment and Analysis

Calculate the accurate concentration of the KMnO$_4$ solution. Calculate and analyze the relative average deviation[37].

[37] 计算并分析相对平均偏差

Safety and Waste Disposal

(1) Wear a lab coat and safety goggles at all times in the laboratory. Wear gloves when handling hot glassware. If any chemicals splash into the eyes or onto the skin, wash with water immediately.

(2) KMnO$_4$ is one of the strongest oxidizing agents available. You MUST be careful when handling it!!!

(3) Never pour the disposed KMnO$_4$ solution down the sink, dispose of it in the designated container. Or treat it first with a reductive substance[38], like excess Na$_2$C$_2$O$_4$ in this experiment.

[38] 先用还原性物质处理

Post-lab Questions

If tap water is used to wash the burette after titration, it will turn brown. Provide a brief explanation why it happened, and write the related chemical reactions.

扫一扫 视频 4-5 KMnO$_4$溶液的配制与标定

Part 2. Preparation and Standardization of 0.1 mol·L^{-1} Na$_2$S$_2$O$_3$ Solution

Objectives

(1) Understand the principle of redox titrations.

(2) Understand the principle of iodimetry and iodometry[39].

[39] 直接碘量法和间接碘量法

(3) Prepare and standardize a $Na_2S_2O_3$ solution.

Principles

Key terms: redox titrations, iodometry, standard $Na_2S_2O_3$ solution.

Since sodium thiosulphate ($Na_2S_2O_3$) is not a primary standard, an indirect method is used to prepare its standard solution. There are two methods to standardize the solution: iodimetry and iodometry. In both methods, a starch solution is used as an indicator.

Iodimetry: Iodine (I_2) is used as the oxidizing agent, typically performed in neutral or slightly acidic condition[40]. This is one of the most important redox titration methods because iodine (I_2) reacts directly, rapidly, and quantitatively[41] with many organic and inorganic substances. It is a direct titration, involving only a single chemical reaction.

$$I_2 + 2S_2O_3^{2-} \rightleftharpoons 2I^- + S_4O_6^{2-}$$

Iodometry: Iodide (I^-) is used as a reducing agent, while oxidizing agents such as potassium iodate (KIO_3) or potassium dichromate ($K_2Cr_2O_7$) serve as the primary standards. The quantitatively oxidizing agent is treated with excess potassium iodide (KI) in an acidic medium, which releases iodine (I_2). The released iodine (I_2) is then titrated with a $Na_2S_2O_3$ solution[42]. This method is an indirect titration, involving at least two chemical reactions.

Step 1.

$$5I^- + IO_3^- + 6H^+ \rightleftharpoons 3I_2 + 3H_2O$$

or

$$8I^- + Cr_2O_7^{2-} + 16H^+ \rightleftharpoons 2Cr^{3+} + 4I_2 + 8H_2O$$

Step 2.

$$I_2 + 2S_2O_3^{2-} \rightleftharpoons 2I^- + S_4O_6^{2-}$$

Therefore

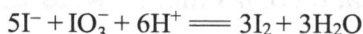
$$IO_3^- \rightarrow 3I_2 \rightarrow 6S_2O_3^{2-}$$

or

$$Cr_2O_7^{2-} \rightarrow 4I_2 \rightarrow 8S_2O_3^{2-}$$

When performing iodimetry or iodometry, follow these guidelines to minimize titration errors[43]:

(1) Avoid a low or high pH.

(2) Titrate quickly while gently swirling at the start.

(3) Add the starch indicator near the endpoint.

(4) After adding the starch indicator, titrate slowly while swirling vigorously[44].

[40] 中性或弱酸性条件

[41] 定量地

[42] 在酸性介质中定量的氧化剂与过量 KI 反应，释放出的 I_2 用 $Na_2S_2O_3$ 滴定

[43] 减小滴定误差

[44] 加入淀粉指示剂后，需慢滴快摇

Pre-lab Questions

How do you minimize titration errors when performing iodometry? Give a brief description.

Apparatus and Reagents

Apparatus: analytical balance, hot plate, burette (25 mL), pipette (20 mL), Erlenmeyer flask (3×150 mL), graduated cylinder, beaker, brown reagent bottle (500 mL).

Reagents: KIO_3 (s, primary standard, $M_r = 214.0$), $K_2Cr_2O_7$ (s, primary standard, $M_r = 294.2$), Na_2CO_3 (s), $Na_2S_2O_3 \cdot 5H_2O$ (s, $M_r = 248.2$), KI (s), H_2SO_4 (3 mol·L^{-1}), starch (0.5%).

Procedures

1. Preparation of 0.1 mol·L^{-1} Na₂S₂O₃ Solution

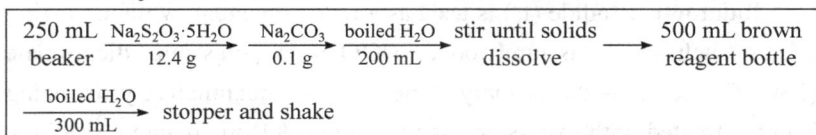

250 mL beaker	$\xrightarrow[\text{12.4 g}]{Na_2S_2O_3 \cdot 5H_2O}$	$\xrightarrow[\text{0.1 g}]{Na_2CO_3}$	$\xrightarrow[\text{200 mL}]{\text{boiled } H_2O}$	stir until solids dissolve	$\longrightarrow$	500 mL brown reagent bottle

$\xrightarrow[\text{300 mL}]{\text{boiled } H_2O}$ stopper and shake

Note: $Na_2S_2O_3$ solution is unstable and should be standardized just before use[45]. For long-term storage, add some Na_2CO_3 to achieve a final concentration of approximately 0.2 g·L^{-1}.

[45] 临用前标定

2. Standardization of 0.1 mol·L^{-1} Na₂S₂O₃ Solution

Method 1: Using KIO₃ as a primary standard.

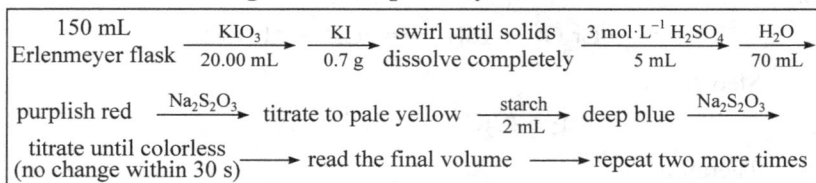

150 mL Erlenmeyer flask	$\xrightarrow[\text{20.00 mL}]{KIO_3}$	$\xrightarrow[\text{0.7 g}]{KI}$	swirl until solids dissolve completely	$\xrightarrow[\text{5 mL}]{3 \text{ mol·L}^{-1} H_2SO_4}$	$\xrightarrow[\text{70 mL}]{H_2O}$

purplish red $\xrightarrow{Na_2S_2O_3}$ titrate to pale yellow $\xrightarrow[\text{2 mL}]{\text{starch}}$ deep blue $\xrightarrow{Na_2S_2O_3}$

titrate until colorless (no change within 30 s) $\longrightarrow$ read the final volume $\longrightarrow$ repeat two more times

Note:

(1) Titrate quickly while gently swirling at the beginning.

(2) Add the starch indicator near the endpoint[46].

[46] 临近终点加入淀粉指示剂

(3) After adding the starch indicator, titrate slowly while swirling vigorously.

Method 2: Using K₂Cr₂O₇ as a primary standard.

(1) Accurately weigh 0.36-0.49 g of potassium dichromate ($K_2Cr_2O_7$, $M_r = 294.2$) to the nearest 0.1 mg[47].

[47] 准确称量,称至 0.1 mg

(2) Place it in a 100 mL beaker, add 30 mL of water, and stir until fully dissolve. Use a glass rod to transfer the solution to a 100 mL volumetric flask.

(3) Rinse the beaker and glass rod 2-3 times, transferring each rinsed solution into the volumetric flask[48]. Finally, dilute the solution to the calibration mark. This forms the standard $K_2Cr_2O_7$ solution.

(4) Pipette 20.00 mL of the standard $K_2Cr_2O_7$ solution into a 250 mL iodine flask[49]. Add about 0.7 g of KI, shaking gently until fully dissolved.

(5) Add 5 mL of 3 mol·L^{-1} H_2SO_4, insert the stopper gently[50], and shake to mix well. Keep the iodine flask in the dark for 30 minutes. The solution will turn reddish brown with purplish-red vapor[51].

(6) Remove the iodine flask from the dark. Rinse the stopper and flask with 100 mL of deionized water to dilute the solution[52].

(7) Titrate immediately with 0.1 mol·L^{-1} $Na_2S_2O_3$ solution until the solution turns yellowish-green.

(8) Add 1 mL of 0.5% starch indicator, the solution will turn deep blue. Continue titrating dropwise until the solution turns bright green[53]. If the color remains unchanged for 30 seconds, the endpoint has been reached. Repeat the titration two more times.

Note:

(1) Due to the large amount of I_2 vapor in the iodine flask, it must be sealed tightly. Before titration, pour in a large volume of water to dilute the solution.

(2) Iodine flasks are specifically used for determining iodine content. Each iodine flask is equipped with a ground glass stopper and a cup-shaped top[54].

Data Treatment and Analysis

Calculate the accurate concentration of the $Na_2S_2O_3$ solution. Calculate and analyze the relative average deviation[55].

Safety and Waste Disposal

(1) Wear a lab coat and safety goggles at all times in the laboratory. Wear gloves when handling hot glassware. If any chemicals splash into the eyes or onto the skin, wash with water immediately.

(2) $K_2Cr_2O_7$ is extremely dangerous to the environment and human health, therefore it is rarely used. Handle it with care.

(3) Never pour the $K_2Cr_2O_7$ waste solution down the sink, dispose of it in the designated container. Or treat first with a reductive substance.

Post-lab Questions

If the following circumstances were to occur during the titration process, analyze their effect on the concentration of $Na_2S_2O_3$ solution using KIO_3 as the primary standard.

[48] 每次将润洗液转移至容量瓶中

[49] 碘量瓶

[50] 轻轻盖紧

[51] 紫红色蒸气

[52] 稀释溶液(减少碘蒸气蒸发)

[53] 亮绿色

[54] 碘量瓶带有磨口塞和杯状瓶口

[55] 计算并分析相对平均偏差

(1) After adding KI, titrate immediately without diluting with water.

(2) Swirl the iodine flask violently at the beginning of titration.

(3) Add the starch too early.

(4) Titrate too slowly at the end of titration.

扫一扫　视频 4-6　Na₂S₂O₃ 溶液的配制与标定

（曾秀琼编写）

Expt. 7　Preparation and Standardization of the Solution for Complexometric Titration
配位滴定溶液的配制与标定

Introduction to Complexometric Titration

[56] 配位化合物是通过配位共价键形成的

[57] 配位滴定(又称络合滴定)

[58] 国际标准与指令

Coordination compounds (also known as complexes) are chemical compounds formed through coordinated covalent bonds[56]. Developed in 1945, complexometric titration (also known as complexation titration)[57] is among the most widely used titrimetric analysis methods and is recommended in many international standards and directives[58]. Complexometric titration is based on complex-forming reaction and is used to determine the concentration of various metal ions in solution.

[59] 乙二胺四乙酸

[60] 由 4 个氧和 2 个氮作为配位原子的六齿配体

EDTA: Ethylenediamine tetra-acetic acid[59] (EDTA, H_4Y) was first synthesized in 1935. It is a hexadentate ligand, consisting of four oxygen and two nitrogen donor atoms[60] (Fig. 4-4).

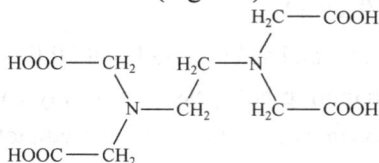

Fig. 4-4　Structure of EDTA

[61] 酸性溶液中 EDTA 有 7 种不同型体

[62] 分布分数

[63] 配位滴定需在缓冲溶液中进行,因为配位反应总释放质子

[64] 只有 EDTA 的 Y^{4-} 型体才能与金属离子配位

Distribution fraction of EDTA. EDTA exists seven different forms in acidic solutions[61]. Fig. 4-5 shows the distribution fraction[62] of EDTA as a function of pH. Complexometric titrations must be performed in buffer solutions because the complexation reaction always liberates protons[63]. **Note:** Only the Y^{4-} form of EDTA can complex with metal cations[64], making pH a critical factor in this titration.

$$H_6Y^{2+} \underset{+H}{\overset{-H}{\rightleftharpoons}} H_5Y^+ \underset{+H}{\overset{-H}{\rightleftharpoons}} H_4Y \underset{+H}{\overset{-H}{\rightleftharpoons}} H_3Y^- \overset{-H}{\underset{+H}{\rightleftharpoons}}$$

$$H_2Y^{2-} \underset{+H}{\overset{-H}{\rightleftharpoons}} HY^{3-} \underset{+H}{\overset{-H}{\rightleftharpoons}} Y^{4-}$$

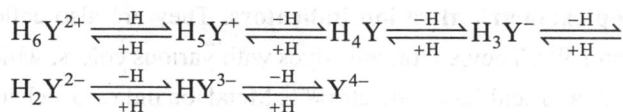

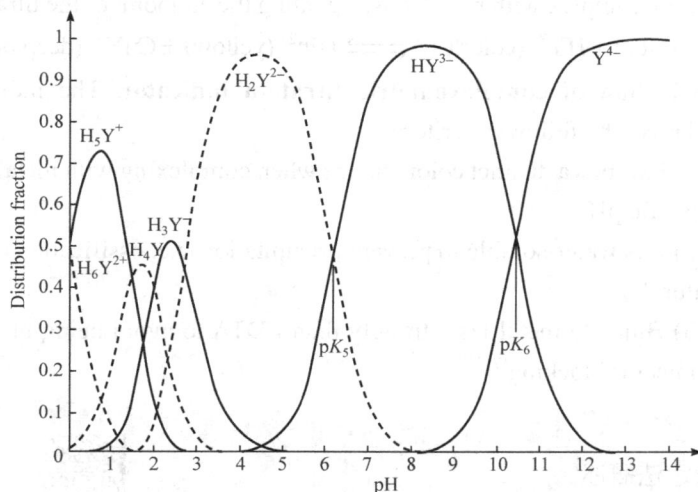

Fig. 4-5　Distribution fraction of EDTA

Characteristics of M-Y complexes.

(1) Stable (chelate, five 5-membered rings[65]) and water-soluble (Fig. 4-6).

[65] (螯合物，5 个五元环)

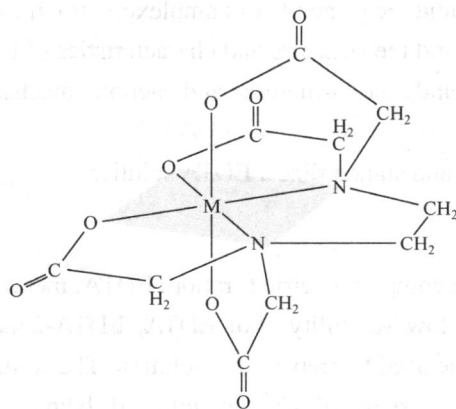

Fig. 4-6　Structure of M-Y chelate

(2) Quantitative and instantaneous with the ratio of 1 : 1[66].

[66] 按 1∶1 与金属离子定量快速配位

(3) Colorless if the metal ion is colorless, or deeper color if the metal ion is colored.

Requirements for an effective complexometric titration[67]. Due to $\Delta pM_{uncertain}$ about 0.2, $c(M)K_f'(MY) \geqslant 10^6$ if $E_T \leqslant \pm 0.1\%$, $\lg K_f'(MY) \geqslant 8$ when $c(M) = 0.01 mol \cdot L^{-1}$.

[67] 有效配位滴定(直接准确滴定)条件

[68] 金属(离子)指示剂

[69] 有机染料, 作为配体和酸碱指示剂

Complexometric titration indicators. They are also called metal ion indicators[68]. They are organic dyes with various colors, which act as both ligands and acid-base indicators[69]. Metal ion indicators change color when they complex with metal ions, signaling the endpoint of the titration.

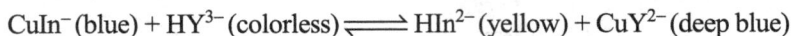

$$CuIn^- (blue) + HY^{3-} (colorless) \rightleftharpoons HIn^{2-} (yellow) + CuY^{2-} (deep blue)$$

Selection of complexometric titration indicator. The indicator should meet the following criteria.

(1) Exhibits a distinct color change when complexing with metal ions at a specific pH.

(2) It is water-soluble to prevent precipitation (i.e. ossification of the indicator[70]).

[70] 防止沉淀(指示剂僵化)

[71] (指示剂封闭)

(3) Binds to metal less strongly than EDTA to avoid being blocked (i.e. indicator blocking[71]).

扫一扫 视频 4-7 配位滴定法概述

Preparation and Standardization of 0.01 mol·L⁻¹ EDTA Solution

Objectives

(1) Understand the principle of complexometric titration.

(2) Understand the structure and characteristics of EDTA.

(3) Understand the structure and action mechanism of metal indicators.

(4) Prepare and standardize a EDTA solution.

Principles

Key terms: complexometric titration, EDTA, metal indicators.

[72] 溶解度低

Due to the low solubility[72] of EDTA, EDTA-2Na [11.1 g·(100 g $H_2O)^{-1}$] should be used to prepare this solution. The solution will then be standardized by a standard Ca^{2+} solution if being used in alkaline system[73], or by a standard Zn^{2+} solution if being used in acidic system.

[73] 若用于碱性体系, 用 Ca²⁺标准溶液标定

Pre-lab Questions

(1) Describe the characteristics of EDTA and M-Y.

(2) Describe the structure and action mechanism of metal ion indicators.

Apparatus and Reagents

Apparatus: analytical balance, top-loading balance, volumetric flask

(100 mL, 250 mL), burette (25 mL), pipette (20 mL), Erlenmeyer flask (3×150 mL), graduated cylinder, beaker, reagent bottle (250 mL).

Reagents: $CaCO_3$ (s, primary standard, $M_r = 134.0$), ZnO (s, primary standard, $M_r = 81.39$), EDTA-2Na (s, $M_r = 336.2$), calconcarboxylic acid (Ca-indicator, s)[74], HCl (6 mol·L^{-1}), NaOH (6 mol·L^{-1}), xylenol orange[75] (0.2%), hexamethylene tetramine (20%).

[74] 钙羧酸(钙指示剂)

[75] 二甲酚橙(指示剂)

Procedures

1. Preparation of 0.01 mol·L^{-1} EDTA Solution

$$\boxed{\begin{array}{l} \underset{\text{beaker}}{250\ mL} \xrightarrow[\text{0.9 g}]{\text{EDTA-2Na}} \xrightarrow[\text{50 mL}]{H_2O} \text{dissolve} \longrightarrow \underset{\text{reagent bottle}}{250\ mL} \xrightarrow[\text{200 mL}]{H_2O} \text{stopper and shake} \end{array}}$$

2. Standardization of 0.01 mol·L^{-1} EDTA Solution

Method 1: Using calcium carbonate ($CaCO_3$) as a primary standard.

(1) Preparation of standard Ca^{2+} solution[76].

[76] (小烧杯底部最好湿润，防止 $CaCO_3$ 粉末被扬起而损失)

$$\boxed{\begin{array}{l} \underset{\text{beaker}}{100\ mL} \xrightarrow[\text{0.10-0.13 g}]{CaCO_3} \xrightarrow[\text{5 mL}]{6\ mol·L^{-1}\ HCl} \text{dissolve} \xrightarrow[\text{30 mL}]{H_2O} \text{mix} \longrightarrow \underset{\text{volumetric flask}}{100\ mL} \\ \xrightarrow{H_2O} \text{dilute to mark} \end{array}}$$

(2) Standardization of 0.01 mol·L^{-1} EDTA solution[77].

[77] (终点为纯蓝色，不夹红色)

$$\boxed{\begin{array}{l} \underset{\text{Erlenmeyer flask}}{150\ mL} \xrightarrow[\text{20.00 mL}]{Ca^{2+}\ sln.} \xrightarrow[\text{25 mL}]{H_2O} \xrightarrow[\text{2.5 mL}]{6\ mol·L^{-1}\ NaOH} \xrightarrow{\text{Ca-indicator}} \\ \text{titrate with EDTA} \longrightarrow \text{from red to blue (no change within 30 s)} \longrightarrow \\ \text{read the final volume} \longrightarrow \text{repeat twice} \end{array}}$$

Method 2: Using zinc oxide (ZnO) as a primary standard.

Using weighing by difference method, accurately weigh 0.15-0.20 g of ZnO to the nearest 0.1 mg[78] in a 100 mL wet beaker. Add 5 mL of 6 mol·L^{-1} HCl to dissolve it completely. Use a glass rod to transfer the solution to a 250 mL volumetric flask. Rinse the beaker and glass rod 2-3 times, each time transferring the rinsed solution into the volumetric flask[79]. Finally, dilute to the mark with deionized water. This forms a standard Zn^{2+} solution.

[78] 差减法称至 0.1 mg

[79] 每次将润洗液转移至容量瓶中

Pipette 20.00 mL of standard Zn^{2+} solution into a 150 mL Erlenmeyer flask. Add 25 mL of water and 1-2 drops of 0.2% xylenol orange as the indicator. Add 20% hexamethylene tetramine dropwise[80] until the solution turns purplish-red, then add an additional 5 mL of 20% hexamethylene tetramine. Titrate with 0.01 mol·L^{-1} EDTA until the solution turns bright yellow[81]. If the color remains unchanged for 30 seconds, the titration endpoint has been reached. Record the volume consumed. Repeat the titration two more times.

[80] 逐滴加入六次甲基四胺

[81] 亮黄色

Data Treatment and Analysis

Calculate the accurate concentration of the EDTA solution. Calculate and analyze the relative average deviation.

Safety and Waste Disposal

(1) Wear a lab coat and safety goggles at all times in the laboratory. If any chemicals splash into the eyes or onto the skin, wash with water immediately.

(2) Because disposed EDTA solution has low environmental impact, it can be safely poured down the sink.

Post-lab Questions

How does ossification of indicator occur? How does indicator blocking occur? Give a brief explanation.

扫一扫　视频 4-8　EDTA 溶液的配制与标定

(曾秀琼编写)

Expt. 8　Preparation and Standardization of the Solution for Precipitation Titration
沉淀滴定溶液的配制与标定

Introduction to Precipitation Titration

1. Some Important Terms

(1) Dissolution and precipitation equilibria: This equilibrium is established when the rates of dissolution and precipitation of a solute are equal[82].

$$A_nB_m(s) \rightleftharpoons nA^{m+}(aq) + mB^{n-}(aq)$$

(2) Solubility product[83]**(K_{sp}):** It is the equilibrium constant for the balance between a solid and its dissolved ions in a given solution.

$$K_{sp} = c^n(A^{m+}) \cdot c^m(B^{n-})$$

(3) Solubility(s): It is the maximum quantity of a substance (A_nB_m) that can be dissolved in a solvent[84].

$$K_{sp} = c^n(A^{m+}) \cdot c^m(B^{n-}) = (ns)^n \cdot (ms)^m$$

[82] 溶解-沉淀平衡：溶质的溶解和沉淀达到相同速率时的平衡

[83] 溶度积

[84] 溶解度：溶质在溶剂中被溶解的最大质量

(4) **Relationship between solubility and solubility product:** For substances with the same ionic structure, a higher solubility product indicates greater solubility. For substances with different ionic structure, first calculate the solubility using the solubility product data, then compare the solubility values[85].

(5) **Predication of precipitate formation:** The ion product (Q_i) of a salt is the product of the concentrations of the ions in solution, raised to the powers specified in the solubility product expression[86]. Q_i is used to predict whether a precipitate will form. There are three possible situations:

1) $Q_i > K_{sp}$: The solution is supersaturated[87], and precipitate will form.

2) $Q_i = K_{sp}$: The solution is saturated and at an equilibrium.

3) $Q_i < K_{sp}$: The solution is unsaturated[88], no precipitate forms, and more solids will dissolve.

2. Requirements for the Precipitation Reactions

Precipitation titration is a type of titration based on a precipitation reaction. These reactions must meet the following criteria:

(1) Take place quickly, without easily forming a supersaturated solution.

(2) Precipitate completely with a stable and low-soluble composition[89].

(3) Detect the titration endpoint well and simply.

(4) Do not cause significant errors due to the absorption ability of precipitates.

3. Argentimetry[90]

Argentimetry is a type of precipitation titration that involves the formation of insoluble silver salts. There are three common argentometry methods, each with a different approach to detecting the titration endpoint.

(1) **Mohr method:** This is one of the oldest titration methods still in use, established by Karl Friedrich Mohr in 1856. In the Mohr method, chromate ion (CrO_4^{2-}) is used as an indicator. After the stoichiometric point, CrO_4^{2-} reacts with excess Ag^+ to form a brick-red precipitate of silver chromate (Ag_2CrO_4) signaling the endpoint[91].

(2) **Volhard method:** Fe(III) is used as an indicator and a standard NH_4SCN solution is used as the titrant. At the endpoint, excess NH_4SCN reacts with Fe(III) to form a red complex ($[FeSCN]^{2+}$), signaling the endpoint. This method is widely used for determining silver (Ag^+) and chloride (Cl^-) ions because the titration can be performed in acidic solutions.

[85] 对于不同离子构型的物质，需先根据溶度积数据计算溶解度，再进行比较

[86] 离子积是将各离子浓度代入溶度积表达式后得到的浓度积

[87] 过饱和溶液

[88] 不饱和溶液

[89] 沉淀完全，具有稳定和溶解度小的组成

[90] 银量法

[91] 化学计量点后，CrO_4^{2-} 与过量 Ag^+ 反应生成 Ag_2CrO_4 砖红色沉淀，指示终点

(3) Fajans method: Developed by Kazimierz Fajans, this method utilizes the indicator absorption technique. At the endpoint, the indicator adheres to the surface of the silver salt precipitate, causing a color change due to the adsorption process[92].

4. Considerations for the Precipitation Titrations

(1) When do the ions begin to precipitate? The answer is $Q_i > K_{sp}$.

(2) When are the ions precipitated completely? The answer is $c_{remaining}$[93] $\leqslant 10^{-6}$ mol·L^{-1}.

(3) How can the mixed ions be precipitated separately and completely[94]? Answer it by yourself.

Preparation and Standardization of 0.04 mol·L^{-1} AgNO₃ Solution

Objectives

(1) Understand the principle of precipitation titration.

(2) Understand the principle of Mohr method.

(3) Prepare and standardize a AgNO₃ solution.

Principles

Key terms: precipitation titration, argentimetry, Mohr method, semimicro-quantitative analysis[95].

If an indirect method is used to prepare the standard AgNO₃ solution, it can be standardized using sodium chloride (NaCl) as a primary standard, with chromate ion (CrO_4^{2-}) serving as the indicator.

$$2Ag^+ + CrO_4^{2-} = Ag_2CrO_4 \downarrow \text{ (brick-red)}$$

The Mohr method has three main requirements.

(1) Control the system's acidity within a pH range of 6.5-10.5.

(2) Control the amount of the indicator.

(3) Do not use Cl$^-$ to titrate a AgNO₃ solution, as Ag₂CrO₄ is difficult to convert into AgCl[96].

Pre-lab Questions

(1) List the requirements of a chemical reaction suitable for precipitation titration. Explain briefly.

(2) List the requirements of the Mohr method. Explain briefly.

Apparatus and Reagents

Apparatus: analytical balance, top-loading balance, volumetric flask (50 mL, brown), burette (10 mL, brown), pipette (10 mL), Erlenmeyer flask (3×150 mL), graduated cylinder, beaker.

[92] 指示剂吸附到银盐沉淀表面，导致指示剂变色

[93] 残留离子浓度

[94] 分步沉淀完全

[95] 半微量定量分析

[96] 不能用 Cl$^-$ 滴定 AgNO₃，因为 Ag₂CrO₄ 难以转化成 AgCl

Reagents: NaCl (s, primary standard, $M_r = 58.44$. Dry at 110°C for about 1 hour before use), AgNO$_3$ (s, $M_r = 169.9$), K$_2$CrO$_4$ (0.1 mol·L^{-1}).

Procedures

Statement:

(1) Due to the high cost of AgNO$_3$, the semimicro-quantitative analytical method is used in this experiment.

(2) AgNO$_3$ solution is light-sensitive, so the amber (brown) reagent bottle and burette are used in this experiment[97].

[97] 采用棕色试剂瓶和滴定管

1. Preparation of 0.04 mol·L^{-1} AgNO$_3$ Solution

100 mL beaker	$\xrightarrow[\text{0.3 g}]{\text{AgNO}_3}$	$\xrightarrow[\text{50 mL}]{\text{H}_2\text{O}}$	stir until solids dissolve completely	$\longrightarrow$ 50 mL brown reagent bottle

Note: All glassware used in this step must be thoroughly cleaned and rinsed several times with deionized water. Otherwise, the AgNO$_3$ solution will be muddy[98].

[98] 否则，AgNO$_3$ 溶液是浑浊的

2. Preparation of 0.04 mol·L^{-1} Standard NaCl Solution

100 mL beaker	$\xrightarrow[\text{0.20-0.25 g}]{\text{NaCl}}$	$\xrightarrow[\text{50 mL}]{\text{H}_2\text{O}}$	stir until solids dissolve completely	$\longrightarrow$ 100 mL volumetric flask
$\xrightarrow{\text{H}_2\text{O}}$ dilute to the mark	$\longrightarrow$ stopper and shake			

3. Standardization of 0.04 mol·L^{-1} AgNO$_3$ Solution[99]

150 mL Erlenmeyer flask	$\xrightarrow[\text{10.00 mL}]{\text{NaCl solution}}$	$\xrightarrow[\text{10 mL}]{\text{H}_2\text{O}}$	$\xrightarrow[\text{1.0 mL}]{\text{K}_2\text{CrO}_4}$	titrate with AgNO$_3$ sln. $\longrightarrow$
until the solution turns pink	$\longrightarrow$ record the volume	$\longrightarrow$ repeat twice		

[99] 为减少 AgNO$_3$ 溶液用量，此处采用 10 mL 滴定管

Data Treatment and Analysis

Calculate the accurate concentration of the AgNO$_3$ solution. Calculate and analyze the relative average deviation.

Safety and Waste Disposal

(1) Wear a lab coat and safety goggles at all times in the laboratory. If any chemicals splash into the eyes or onto the skin, wash with water immediately.

(2) Because K$_2$CrO$_4$ is extremely dangerous to the environment and human health, do not pour K$_2$CrO$_4$ waste solution down the sink, dispose of it in the designated container.

(3) AgNO$_3$ is an expensive chemical, so pour the excess AgNO$_3$ solution into the designated container for reuse.

Post-lab Questions

Discuss the determination error if the indicator amount is less or higher.

扫一扫 视频 4-9 沉淀滴定法概述

(曾秀琼编写)

Expt. 9 Practise of Basic Operations for Titrimetric Analysis
滴定分析基本操作训练

Objectives

[100] 指示剂的作用机理

(1) Master the theory of titrimetric analysis and acid-base titration.
(2) Understand the action mechanism of indicator[100].
(3) Practice the basic operations of titrimetric analysis.

Principles

Key terms: primary standard, acid-base titration, indicator.

Acid-base titration is an important method in titrimetric analysis. Please refer to Expt.5 for more details about the principles.

Since hydrochloric acid (HCl) is a strong monoprotic acid and sodium hydroxide (NaOH) is a strong monoprotic base, their molar ratio is 1 : 1, and the stoichiometric point of the reaction of HCl with NaOH is pH = 7.0[101].

[101] 一元强酸和一元强碱反应的化学计量点 pH = 7.0

$$NaOH + HCl == NaCl + H_2O$$
$$c_{HCl} \times V_{HCl} = c_{NaOH} \times V_{NaOH}$$

When titrating HCl with NaOH, the pH color change interval is 4.3-9.7, so phenolphthalein (pH range: 8.3-10.0, colorless→pink) is used as an indicator[102].

[102] 用 NaOH 滴定 HCl 时，以酚酞为指示剂

When titrating NaOH with HCl, the pH color change interval is 9.7-4.3, so methyl orange (pH range: 3.1-4.4, yellow→orange) or methyl red (pH range: 4.4-6.2, orange→red) is used as an indicator[103]. **Note:** Methyl orange is typically preferred because its color change at the endpoint is easier to detect.

[103] 用 HCl 滴定 NaOH 时，以甲基橙或甲基红为指示剂

Pre-lab Questions

(1) Watch the on-line teaching videos, briefly summarize the basic

operations of an analytical balance, volumetric flask, burette, and volumetric pipette. Submit the assignments online.

(2) Solve the two pre-lab questions in Expt. 5.

Apparatus and Reagents

Apparatus: analytical balance, hot plate, burette (25 mL), volumetric flask (100 mL), pipette (20 mL), Erlenmeyer flask (3×150 mL), graduated cylinder, beaker.

Reagents: potassium hydrogen phthalate[104] (s, primary standard, $KHC_8H_4O_4$, $M_r = 204.2$), NaOH (s), HCl (6 mol·L^{-1}), phenolphthalein (0.1%), methyl orange (0.1%).

[104] 邻苯二甲酸氢钾(基准物质)

Procedures

1. Preparation of Two Solutions

For more details, refer to Expt. 5. These prepared solutions can be shared by two people.

(1) Preparation of 0.1 mol·L^{-1} NaOH solution.

Prepare 250 mL NaOH solution in a reagent bottle, ensuring it is sealed with a rubber stopper.

(2) Preparation of 0.1 mol·L^{-1} HCl solution.

Prepare 250 mL HCl solution and use 6 mol·L^{-1} HCl stock solution in this step.

2. Preparation of Volumetric Glassware[105]

For more details, refer to Expt. 5.

(1) Clean and rinse two burettes. To avoid confusion, use one burette with a red stopcocker and another with a blue stopcocker[106].

(2) Clean and rinse one pipette[107].

(3) Clean one volumetric flask[108].

[105] 容量玻璃器皿

[106] 两支滴定管旋塞分别是红色和蓝色

[107] 清洗并润洗移液管

[108] 容量瓶

3. Practise of Basic Operations for Titrimetric Analysis

(1) Titrating NaOH with HCl.

150 mL Erlenmeyer flask $\xrightarrow{0.1\ mol·L^{-1}\ NaOH\ 20.00\ mL}$ $\xrightarrow{methyl\ orange\ 1\text{-}2\ drops}$ $\xrightarrow{titrate\ with\ 0.1\ mol·L^{-1}\ HCl}$ from yellow to orange (no change within 30 s) $\longrightarrow$ record the final volume $\longrightarrow$ repeat two more times

Note: Drain about 20 mL (to the nearest 0.01mL) NaOH from the burette into a 150 mL Erlenmeyer flask[109]. Or pipette 20.00 mL NaOH from the reagent bottle into a 150 mL Erlenmeyer flask[110].

[109] 从滴定管准确放出

[110] 用移液管移取

(2) Titrating HCl with NaOH.

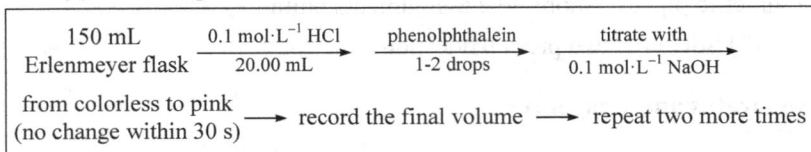

150 mL Erlenmeyer flask	$\xrightarrow[\text{20.00 mL}]{0.1 \text{ mol·L}^{-1} \text{ HCl}}$	phenolphthalein 1-2 drops	$\xrightarrow{}$	titrate with 0.1 mol·L^{-1} NaOH

from colorless to pink (no change within 30 s) $\xrightarrow{}$ record the final volume $\xrightarrow{}$ repeat two more times

Note: Drain about 20 mL (to the nearest 0.01mL) HCl from the burette into a 150 mL Erlenmeyer flask. Or pipette 20.00 mL HCl from the reagent bottle into a 150 mL Erlenmeyer flask.

(3) Standardization of 0.1 mol·L^{-1} NaOH solution.

For more details, refer to Expt. 5.

Data Treatment and Analysis

[111] 计算并分析相对平均偏差

(1) Calculate the ratios of NaOH to HCl and HCl to NaOH separately. Calculate and analyze the relative average deviation[111].

(2) Calculate the accurate concentration of NaOH. Calculate and analyze the relative average deviation.

Safety and Waste Disposal

(1) Wear a lab coat and safety goggles at all times in the laboratory. If any chemicals splash into the eyes or onto the skin, wash with water immediately.

(2) NaOH and HCl are corrosive, they must be handled with care. Wear gloves whenever needed.

(3) Do not pour the remaining acid or base solutions down the sink. First neutralize them with each other.

Post-lab Questions

If the concentrations of these two solutions are the same, and ignoring any operational errors, the determined ratio of NaOH to HCl is different from the ratio of HCl to NaOH. How did this happen? Give a brief explanation.

扫一扫　视频 4-10　酸碱滴定练习

(蔡吉清　曾秀琼编写)

Expt. 10　Determination of the Molecular Weight of an Unknown Diprotic Organic Acid
未知二元有机酸分子量的测定

Objectives

(1) Understand the principle of polyprotic acid-base titrations[112].

(2) Practice the basic operations of titrimetric analysis[113].

(3) Determine the molecular weight of an unknown acid using acid-base titration.

Principles

Key terms: primary standard[114], polyprotic acid-base titration, indicator.

The unknown in this experiment is a diprotic organic acid[115] (H$_2$A, $K_{a1} = 5.9\times10^{-2}$, $K_{a2} = 6.4\times10^{-5}$). The molecular weight[116] (also known as molar mass) of this unknown acid can be determined by titrating it with a standard NaOH solution. The pH at the stoichiometric point is about 9, so phenolphthalein[117] (PP. pH range: 8.3-10.0, colorless→pink) is used as the indicator. For more details about principles of this experiment, refer to Expt. 5.

$$2NaOH + H_2A \Longrightarrow Na_2A + 2H_2O$$

The precise (accurate) apparatus used in titrimetric analysis include the analytical balance, burette, volumetric flask, and pipette[118]. Proper operations of this volumetric glassware is essential for achieving high accuracy. For more details, refer to Chapter 2.

Pre-lab Questions

(1) Solve the two pre-lab questions in Expt. 5.

(2) The two protons in this diprotic organic acid cannot be titrated stepwise[119], they can only be titrated together. Give a brief explanation.

Apparatus and Reagents

Apparatus: analytical balance, burette (25 mL), volumetric flask (100 mL), pipette (20 mL), Erlenmeyer flask (3×150 mL), graduated cylinder, beaker.

Reagents: potassium hydrogen phthalate[120] (s, primary standard, M_r = 204.2), unknown organic acid (s), NaOH (0.1 mol·L^{-1}), phenolphthalein (0.1%).

[112] 多元酸碱滴定

[113] 滴定分析

[114] 基准物质

[115] 二元有机酸

[116] 分子量

[117] 酚酞

[118] 滴定分析中的精密仪器有分析天平、滴定管、容量瓶和移液管

[119] 两个质子不能被分步滴定

[120] 邻苯二甲酸氢钾(基准物质)

Procedures

1. Preparation of the Sample Solution

```
100 mL    unknown acid    H₂O    stir to dissolve completely         100 mL
beaker  ────────────→  ────────→                           ────→  volumetric flask
           0.5-0.6 g     30 mL

────→ dilute to the mark
```

2. Standardization of 0.1 mol·L⁻¹ NaOH Solution

Refer to Expt. 5. for more details.

3. Determination of the Molecular Weight of the Unknown Organic Acid

```
150 mL        sample solution     phenolphthalein       titrate with
Erlenmeyer flask ───────────→  ──────────────→  ───────────────────→
                  20.00 mL          1-2 drops       0.1 mol·L⁻¹ NaOH

from colorless to pink
(no change within 30 s) ──→ record the volume ──→ repeat two more times
```

Note:

(1) Do not swirl the Erlenmeyer flask vigorously during titration, as the NaOH solution can absorb CO_2 from the air[121].

(2) After reaching the endpoint, avoid swirling the Erlenmeyer, as the pink solution will quickly turn colorless.

[121] 滴定过程中不要剧烈摇动锥形瓶,因为 NaOH 溶液会吸收 CO_2

Data Treatment and Analysis

(1) Calculate the accurate concentration of the NaOH solution. Calculate and analyze the relative average deviation.

(2) Calculate the molecular weight of the unknown diprotic organic acid and estimate its identity. **Note:** Each hydrate molecule contains two molecules of water[122].

[122] 每个水合物分子含两分子结晶水

Safety and Waste Disposal

(1) Wear a lab coat and safety goggles at all times in the laboratory. If any chemicals splash into the eyes or onto the skin, wash with water immediately.

(2) NaOH solid and solution are corrosive, they must be handled with care. Wear gloves whenever needed.

(3) Do not pour any disposed acidic or alkaline solution down the sink. First neutralize them with each other[123].

[123] 相互中和

Post-lab Questions

How can the two protons in this diprotic organic acid be titrated stepwise?

Design an experiment procedure.

Exploring Experiment

There is an unknown industrial material with a purity of 95%. It should be one of the following two diprotic organic acids, acid-A (M_r = 150.0, $K_{a1} = 1.0 \times 10^{-3}$, $K_{a2} = 4.6 \times 10^{-5}$) or acid-B ($M_r$ = 138.1, K_{a1} =1.0 × 10^{-3}, $K_{a2} = 4.2 \times 10^{-13}$). Design an experimental method to find out which one it is, and determine its accurate purity.

The following reagents are available: unknown organic acid (s), standard NaOH solution (0.1 mol·L^{-1}), phenolphthalein (0.1%), methyl orange (0.1%).

扫一扫   视频 4-11　未知酸分子量的测定

（蔡吉清　曾秀琼编写）

Chapter 5　Comprehensive and Designing Experiments
综合设计实验

Expt. 11　Determination of the Fluoride Content in Tea Samples by FISE
氟离子选择电极法测定茶叶中氟离子含量

Objectives

[1] 电位分析法

(1) Understand the principles of potential analysis[1].

(2) Learn the construction and action mechanism of fluoride ion selective electrodes[2].

[2] 氟离子选择电极

[3] 标准曲线法

(3) Use standard curve method[3] and standard addition method[4] to determine the fluoride content in tea samples.

[4] 标准加入法

Principles

[5] 膜电极

Key terms: potentiometric analysis, membrane electrode[5], ion selective electrode (ISE), fluoride ion selective electrode (FISE), standard curve method, standard addition method, total ionic strength adjustment buffer[6] (TISAB).

[6] 总离子强度调节缓冲液

[7] 氟元素

Fluorine[7] (F) is an essential element found in various environmental, clinical, and food samples. While small amounts of fluorine intake are beneficial for health, particularly for strong bones and teeth. However, excess intake can be harmful, leading to fluorosis of bone, dental fluorosis[8], or even brain damage. For adults, the lethal dose ranges from 0.20-0.35 g fluorine per kg of body weight.

[8] 氟骨病和氟斑牙

[9] 氟离子

Fluoride[9] (F$^-$) in tea is a significant sources of fluorine (F) intake, as tea plants absorb fluoride from the soil. To minimize the risk of fluoride toxicity from tea, it's recommended to drink teas from young leaves, such as white tea, which contain high levels of antioxidants[10] and lower levels of fluoride. For example, flower teas contain nearly 0 $\mu g \cdot g^{-1}$ of fluoride, while dark teas have the highest content, averaging (296.14 $\pm$ 246.07) $\mu g \cdot g^{-1}$.

[10] 抗氧化剂

[11] 电分析化学

Potentiometry is an electroanalytical chemistry[11] method used to measure the potential difference between two electrodes to determine the concentration of a solute in solution[12]. Its equipment is very simple, including an indicator electrode (usually an ion selective electrode, ISE),

[12] 测定两个电极间的电位(电势), 得到溶质的浓度

a reference electrode, and a potentiometer[13]. The potential of ISE changes with the activity of the analyte[14], while the reference electrode maintains a fixed potential.

ISE is also known as a membrane electrode, is highly selective for specific ions. It typically consists of an internal reference electrode and a sensitive membrane[15] that acts as the interface between the analyte solution and the electrode. The potential across the membrane is determined by the difference in ion activity on each side of the semi-membrane[16].

reference electrode ‖ analyte solution | ion selective electrode

The fluoride ion selective electrode (FISE), developed in 1966, is a type of ion selective electrode sensitive to F^- concentration. In a typical FISE, the sensing element is a crystal membrane made of LaF_3 doped with EuF_2[17] (Fig. 5-1). The internal F^- solution has a fixed activity, so the potential across the membrane is related to the F^- activity in the sample solution[18]. The relationship between the measured potential (E) and the F^- activity (a_{F^-}) is expressed by Equation (5-1).

$$E = K + \frac{2.303RT}{nF}\lg a_{F^-} = K - 0.0592\lg a_{F^-} \quad (5\text{-}1)$$

where R is molar gas constant with the value of 8.314×10^{-3} kJ·mol^{-1}·K^{-1}, T is the thermodynamic temperature[19] (K), F is Faraday constant with the value of 96500 C·mol^{-1}, a is the activity of the analyte ion[20], n represents the charge[21] of the analyte ion. **Note:** n is negative for anions[22], e.g. -1 for F^-.

[13] 指示电极、参比电极和电位计

[14] 离子选择电极的电位随分析物活度改变

[15] ISE 由内参比电极和敏感膜组成

[16] 通过半透膜两侧离子活度之差测定膜电位

[17] (氟电极)敏感膜由氟化镧单晶掺杂少量氟化铕制成

[18] 膜内 F$^-$溶液具有固定的活度，因此膜两侧电位与待测试液的 F$^-$活度有关

[19] 热力学温度

[20] 试液离子的活度

[21] 电荷数

[22] 阴离子的 n 为负数

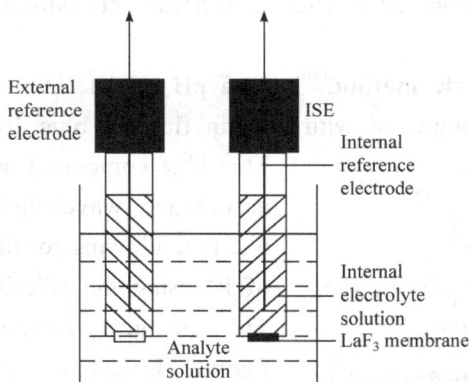

Fig. 5-1   FISE

We know that a equals $c·\gamma$, where γ is the activity coefficient[23]. Equation (5-1) can be transferred into Equation (5-2) when γ remains constant.

$$E = K - 0.0592\lg\gamma - 0.0592\lg c_{F^-} \quad (5\text{-}2)$$

[23] 活度系数

[24] (标准曲线法): 假设
溶液的总离子强度不变(γ
固定), 因此电位(E)与
$\lg c_{\mathrm{F}^-}$ 呈线性关系

Standard curve method. A standard curve represents the relationship between two quantities, allowing the determination of the concentration of an unknown substance. Provided the total ionic strength of the solution remains constant (i.e., γ is fixed), the potential is linearly related to $\lg c_{\mathrm{F}^-}$ [24]. This means the E-pF relationship forms a straight line, enabling the c_{F^-} to be measured using the standard curve [Equation (5-3)].

$$E = K' - 0.0592\lg c_{\mathrm{F}^-} = K' + 0.0592\mathrm{pF} \tag{5-3}$$

[25] (标准加入法): 在待
测试液中加入定量的标
准溶液, 测定加入前后的
电位变化(ΔE)

Standard addition method. When the sample contains unknown or complex components, the standard addition method is preferred. In this method, a known amount of standard F$^-$ solution is added to the analyte, and the potential is measured before and after the addition[25]. Equation (5-4) explains how this method works.

$$c_x = \frac{c_s V_s}{V_x}(10^{\Delta E/S} - 1)^{-1} \tag{5-4}$$

where S is the slope of the ISE, it can be obtained from the slope of the standard curve. c_x and V_x are the concentration and volume of the analyte before addition. c_s and V_s are the concentration and volume of the standard solution added. ΔE is the potential difference before and after addition. **Note:** To minimize the measurement error, c_s should be high, V_s should be small[26], ΔE should be larger than 20 mV.

[26] 为减小测定误差, 加
入的标准溶液需浓度高、
体积小

[27] 总离子强度调节缓
冲液

[28] 柠檬酸

[29] 乙酸缓冲液

[30] 比色法

[31] 茜素氟蓝

[32] 最大吸收波长

Total ionic strength adjustment buffer[27] **(TISAB).** To accurately measure the F$^-$ concentration, a TISAB solution is added to control the acidity and ionic strength of the sample solution. The TISAB used in this experiment contains citric acid[28], an acetic acid buffer[29], and a NaCl solution.

Colorimetric method[30]**.** At a pH of 4.1, F$^-$ can form a blue coordination compound with alizarin fluorine blue[31] and La(NO$_3$)$_3$. This blue compound has a maximum absorbance wavelength[32] ($\lambda_{\max}$) of 610 nm, allowing for the measurement of F$^-$ using the colorimetric method. The chemical formula of alizarin fluorine blue is C$_{19}$H$_{15}$NO$_8$, and Fig. 5-2 shows its structure.

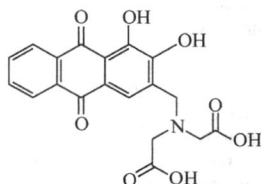

Fig. 5-2   Alizarin fluorine blue

Pre-lab Questions

(1) Briefly describe the principles of potentiometric analysis and ISE.

(2) The TISAB solution used in this experiment has several components. Write out the components and their functions[33].

[33] 组成及作用

(3) Briefly describe the difference between the standard curve method and the standard addition method. Give their advantages and disadvantages.

Apparatus and Reagents

Apparatus: electronic balance (0.01 g), pH meter, magnetic stirrer[34], combination fluoride ion selective electrode[35] (combination FIES), pipette, beaker, colorimetric tube[36], plastic beaker, funnel.

Reagents: standard F^- solution (1.00×10^{-1} mol·L^{-1}). TISAB (dissolve 58 g NaCl and 10 g citric acid in 800 mL deionized water, then add 57 mL acetic acid, and use 40% NaOH to adjust the pH = 5. Finally dilute to the 1 L mark and mix well), tea samples (white tea, green tea, black tea, brick tea[37], etc.).

[34] 磁力搅拌器
[35] 复合氟离子选择电极
[36] 比色管

[37] 砖茶

Procedures

1. Preparation of the Tea Solution[38]

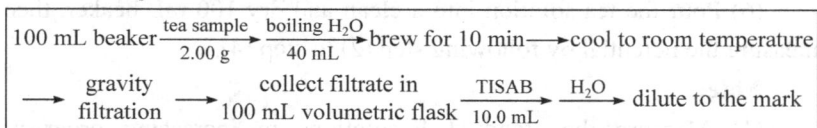

[38] 要点：冷却后常压过滤，润洗茶叶和玻璃棒数次，每次将润洗液转移到漏斗上。可加几滴乙醇消除滤液的泡沫。不可加水超过刻度线

100 mL beaker $\xrightarrow[\text{2.00 g}]{\text{tea sample}}$ $\xrightarrow[\text{40 mL}]{\text{boiling H}_2\text{O}}$ brew for 10 min $\longrightarrow$ cool to room temperature

$\xrightarrow{\text{gravity filtration}}$ collect filtrate in 100 mL volumetric flask $\xrightarrow[\text{10.0 mL}]{\text{TISAB}}$ $\xrightarrow{\text{H}_2\text{O}}$ dilute to the mark

Note:

(1) Rinse the beaker and tea leaves several times, transferring the rinsed solution to the funnel each time.

(2) If foam forms in the solution, add a few drops of alcohol.

(3) Be careful not to add water above the mark when diluting.

2. Preparation of the Standard F^- Solutions[39]

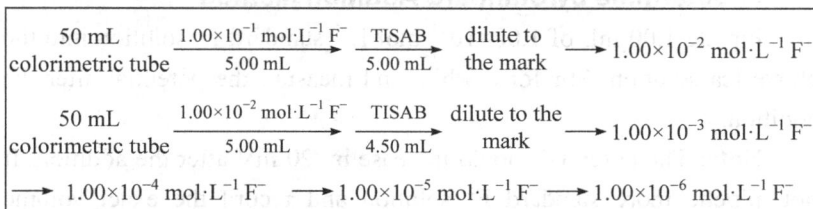

[39] 要点：依次配制 5 份氟离子标准溶液。从第二份起，只需加入 4.50 mL TISAB

50 mL colorimetric tube $\xrightarrow[\text{5.00 mL}]{1.00 \times 10^{-1}\,\text{mol·L}^{-1}\,F^-}$ $\xrightarrow[\text{5.00 mL}]{\text{TISAB}}$ dilute to the mark $\longrightarrow$ 1.00×10^{-2} mol·L^{-1} F^-

50 mL colorimetric tube $\xrightarrow[\text{5.00 mL}]{1.00 \times 10^{-2}\,\text{mol·L}^{-1}\,F^-}$ $\xrightarrow[\text{4.50 mL}]{\text{TISAB}}$ dilute to the mark $\longrightarrow$ 1.00×10^{-3} mol·L^{-1} F^-

$\longrightarrow$ 1.00×10^{-4} mol·L^{-1} F^- $\longrightarrow$ 1.00×10^{-5} mol·L^{-1} F^- $\longrightarrow$ 1.00×10^{-6} mol·L^{-1} F^-

Note:

(1) Prepare five standard F^- solutions, one at a time.

(2) Add 5.00 mL of TISAB to the first standard F^- solution, and add 4.50 mL of TISAB to each of the remaining four solutions.

3. Preparation of Fluoride Ion Selective Electrode (FISE)[40]

[40] 要点：将 pH 计模式设成 mV,放入去离子水和搅拌子，反复用去离子水洗涤复合氟电极，至空白电位大于 360 mV

Connect a combination FISE to a pH meter and set the meter to the mV mode. Add 100 mL of deionized water and a stir bar in a 250 mL beaker, then place the beaker on a magnetic stirrer. Immerse the combination FISE in the water, stir for a while, and then replace it with 100 mL of fresh deionized water, stir again. Repeat this process until the blank potential reaches above 360 mV.

Note: Ensure that the electrode does not touch the bottom of the beaker, and the stir bar does not touch the electrode.

4. Determine by Standard Curve Method

(1) Pour all of the 1.00×10^{-6} mol·L^{-1} standard F$^-$ solution into a clean, dry 50 mL beaker.

(2) Rinse the combination FISE with deionized water and dry it with absorbent paper[41].

[41] 用吸水纸擦干

(3) Place the beaker on the magnetic stirrer plate, add a stir bar, and immerse the combination FISE in the solution.

(4) Stir the solution at a constant speed[42] for 1 minute. Stop stirring and record the potential once it stabilizes.

[42] 匀速搅拌

(5) Repeat the above steps for each remaining F$^-$ standards.

(6) Pour the tea solution into a clean and dry 100 mL beaker, then measure the potential by following step (2) to step (4).

Note:

(1) Measure the standard F$^-$ solutions in increasing order of concentration[43]. Dilute solutions may require more time to reach equilibrium.

[43] 按 F$^-$ 标准溶液浓度由低到高测定

(2) Always rinse and dry the combination FISE between measurements.

(3) Ensure that the standard solutions and tea solution are stirred at the same constant speed.

5. Determine by Standard Addition Method

Pipette 1.00 mL of 1.00×10^{-1} mol·L^{-1} standard F$^-$ solution into the above tea solution. Stir for a while and measure the potential after the addition.

Note: The potential should increase by 20 mV after the addition. If not, pipette more standard F$^-$ solution and record the exact volume added[44].

[44] 记录加入的准确体积

6. Determine by Colorimetric Method

(1) Pipette 10.00 mL of tea solution into a 25 mL colorimetric tube.

(2) Add 10.0 mL of the mixed chromogenic[45], mix well, and let it react for 30 minutes.

[45] 混合显色剂

(3) Dilute the solution to the calibration mark with deionized water.

(4) Compare the resulting color with the color scales[46] made from standard F$^-$ solutions.

[46] 与色阶对比

(5) Determine the concentration range of the tea sample.

Note:

(1) Since the color of tea solution can interfere with the determination, use the white tea in this step.

(2) A spectrophotometer can be used for more accurate measurement result[47].

[47] 可用分光光度计得到更准确的测定结果

Data Treatment and Analysis

(1) According to the data of Table 5-1, use software to plot an E-pF standard curve and determine the linear equation and R^2 of the best fit line[48], then calculate the F$^-$ concentration in the tea sample ($\mu g \cdot g^{-1}$).

[48] 得到线性方程和最佳拟合线的 R^2

Table 5-1　Potentials of the standard F$^-$ solutions and tea sample

Trial	1	2	3	4	5	tea sample
pF						
E/mV						

(2) Substitute the slope of the standard curve into Equation (5-4)[49] to calculate the F$^-$ concentration in the tea sample ($\mu g \cdot g^{-1}$) determined by the standard addition method.

[49] 将标准曲线的斜率代入式(5-4)

$V_x =$____mL, $c_s =$____mol$\cdot$L^{-1}, $V_s =$____mL, $\Delta E =$____mV.

Safety and Waste Disposal

(1) Wear a lab coat and safety goggles at all times in the laboratory. Wear gloves when handling hot glassware. If any chemicals splash into the eyes or onto the skin, wash with water immediately.

(2) High fluoride intake can be dangerous for health. Be careful when handling fluoride solutions.

(3) Pour all solid and liquid waste into their designated containers. Do not pour the tea leaves into the sink.

Clean up. Turn off the pH meter. Rinse the electrode. Hang the electrode in the electrode-handler and leave it to dry[50].

[50] 挂在电极架上晾干

Post-lab Questions

Analyze the results of the two determination methods. Give an explanation.

Exploring Experiment

Design an experiment to measure the concentration of fluoride in some everyday products, such as a toothpaste, a tablet, etc.

扫一扫   视频 5-1   氟离子选择电极测定茶叶中氟含量(讲解)
            视频 5-2   氟离子选择电极测定茶叶中氟含量(实验)

(曾秀琼编写)

Expt. 12　Preparation and Characterization of Mohr's Salt
莫尔盐的制备及表征

Objectives

[51] 复盐

[52] 莫尔盐(又称摩尔盐)

[53] 高锰酸钾滴定法

(1) Understand the concept and characteristics of double salt[51].

(2) Prepare and characterize Mohr's salt[52].

(3) Understand the principles and procedures of permanganate titration[53].

Principles

Key terms: double salt, Mohr's salt, permanganate titration.

[54] 硫酸亚铁铵

Mohr's salt [$(NH_4)_2SO_4·FeSO_4·6H_2O$], also known as ferrous ammonium sulfate[54], is a light green or blue-green crystalline compound. It is named after German chemist Karl Friedrich Mohr, who introduced several important titration methods in the 19th century. Mohr's salt is soluble in water but insoluble in several organic solvents. Unlike ferrous

[55] 不易结块和氧化

[56] 基准物质

sulfate ($FeSO_4$), it does not cake and it is less prone to oxidization[55], making it a primary standard[56]. Mohr's salt is also widely used in industrial applications, such as iron plating.

[57] 复盐由阳离子不同
(不能为 H^+)而阴离子相
同的两种单盐组成，结构
不同于单盐

[58] 溶解度小于两种单盐

A double salt is a crystalline compound that forms from a mixture of two simple salts with different cations but the same anion, and it has a distinct crystal structure compared to either of the simple salt[57]. The solubility of a double salt is lower than either of the simple salt[58], as shown in Table 5-2. Therefore, ferrous ammonium sulfate crystals can be obtained through evaporation and cooling of its solution.

Table 5-2　Solubility of some related substances　　unit: [g·(100 g water)$^{-1}$]

$T/°C$	10	20	30	70
$FeSO_4$ ($M_r = 151.9$)	20.5	26.6	33.2	56.0
$(NH_4)_2SO_4$ ($M_r = 132.1$)	73.0	75.4	78.1	91.9
$(NH_4)_2SO_4·FeSO_4·6H_2O$ ($M_r = 392.1$)	18.1	21.2	24.5	38.5

There are several methods to prepare Mohr's salt. In this experiment, iron powder is first reacted with diluted sulfuric acid[59] (H_2SO_4) to form a $FeSO_4$ solution. Then, ammonium sulfate $(NH_4)_2SO_4$ is added to form a double salt solution. Finally, the solution is evaporated and cooled to obtain crystals.

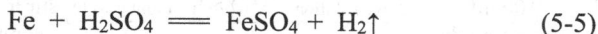

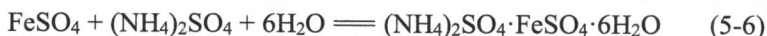

[59] 稀硫酸

$$Fe + H_2SO_4 == FeSO_4 + H_2\uparrow \quad (5\text{-}5)$$

$$FeSO_4 + (NH_4)_2SO_4 + 6H_2O == (NH_4)_2SO_4 \cdot FeSO_4 \cdot 6H_2O \quad (5\text{-}6)$$

The iron content in the product is determined by permanganate ($KMnO_4$) titration, which is a self-catalyst and self-indicator redox titration. When using sodium oxalate ($Na_2C_2O_4$) to standardize the $KMnO_4$ solution, it is important to control the temperature, acidity, and titration speed[60]. For more details, refer to Exp. 6.

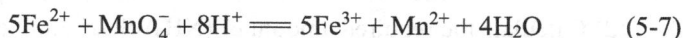

[60] 高锰酸钾滴定法是自身催化和自身指示剂。用 Na₂C₂O₄标定时需注意三个度：温度、酸度和滴定速度

$$5Fe^{2+} + MnO_4^- + 8H^+ == 5Fe^{3+} + Mn^{2+} + 4H_2O \quad (5\text{-}7)$$

$$2MnO_4^- + 5C_2O_4^{2-} + 16H^+ == 2Mn^{2+} + 10CO_2\uparrow + 8H_2O \quad (5\text{-}8)$$

Pre-lab Questions

(1)-(2) Solve the two pre-lab questions of Part 1 in Expt. 6.

(3) When using permanganate titration to determine the iron content in the product, why should the product solution not be heated before titration?

Apparatus and Reagents

Apparatus: hot plate, vacuum pump, suction flask, Buchner funnel, beaker (500 mL, 250 mL, 100 mL), Erlenmeyer flask (3 × 150 mL), volumetric flask (100 mL), burette (25 mL), pipette (20 mL, 10 mL), graduated cylinder[61] (50 mL, 10 mL), evaporating dish[62], watch glass[63].

Reagents: Iron powder(s), $Na_2C_2O_4$(s), $(NH_4)_2SO_4$(s), H_2SO_4 (3 mol·L^{-1}), $KMnO_4$ (0.02 mol·L^{-1}), ethanol.

[61] 量筒
[62] 蒸发皿
[63] 表面皿

Procedures

1. Preparation of FeSO₄

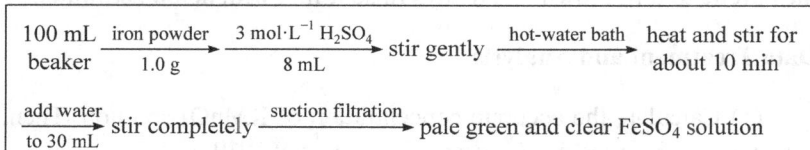

Note:

(1) If a significant amount of white or pale green solids form in the beaker before suction filtration, add a little more water to dissolve them[64].

[64] 抽滤前若有固体析出，补加点水使其溶解

[65] 若滤液不澄清,可再次过滤

[66] 控制溶液总体积,否则蒸发需很长时间

(2) If the filtrate is not clear, repeat the suction filtration process[65].

(3) Ensure the total volume is controlled to 30 mL, as exceeding this volume will prolong the evaporation time[66].

2. Preparation of $(NH_4)_2SO_4 \cdot FeSO_4 \cdot 6H_2O$

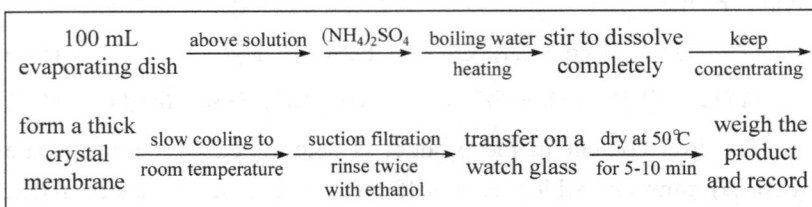

Note:

[67] 沸水浴

(1) Pour about 180 mL of deionized water into a 250 mL beaker, and heat it on a hot plate to prepare a boiling water bath[67].

(2) Calculate the required amount of $(NH_4)_2SO_4$ before class.

[68] 浓缩时不要搅拌溶液,否则析出晶体很小

(3) Avoid stirring the solution while concentrating to prevent the formation of small, sandy crystals[68].

[69] 用乙醇洗涤产品时,抽滤瓶中会析出大量固体,因为溶液(滤液)极性降低了

(4) Rinse the product twice with ethanol, using 5 mL each time. A large number of precipitates will form after rinsing due to the decreased polarity of the solution[69].

3. Standardization of 0.02 mol·L^{-1} KMnO$_4$ Solution

For more details, refer to Expt. 6.

4. Determination of the Iron Content in the Product

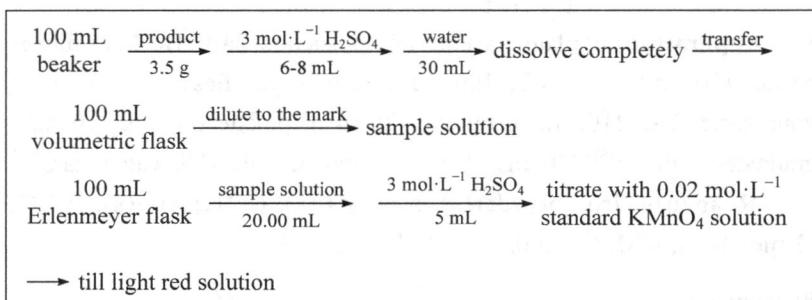

[70] 不能用自来水洗涤滴定管,否则会(因 MnO$_2$ 析出)变褐色

Note: Do not wash the burette directly with tap water, as it may cause the burette to turn brown[70]. Always rinse it first with deionized water.

Data Treatment and Analysis

[71] 相对平均偏差

[72] 测定的相对误差。提示,需先计算莫尔盐中铁含量的理论值

(1) Calculate the accurate concentration of KMnO$_4$ solution. Then, calculate and analyze the relative average deviation[71].

(2) Calculate the iron content and the purity of the product. Calculate and analyze the relative average deviation, as well as the relative error of measurement[72].

Safety and Waste Disposal

(1) Wear a lab coat and safety goggles at all times in the laboratory. Wear gloves when handling the hot glassware. If any chemicals splash into the eyes or onto the skin, wash with water immediately.

(2) $KMnO_4$ solution is highly corrosive with the strong oxidizing ability[73], handle with care.

[73] 强腐蚀性和氧化能力

(3) Never dispose of $KMnO_4$ solution down the sink. Dispose of it to the designated container, or treat it with a reductive substance like excess $Na_2C_2O_4$ or product in this experiment[74].

[74] 将 $KMnO_4$ 废液用剩余的还原性物质处理

Post-lab Questions

How can all the ions in the product be identified using qualitative analysis[75] methods? Write out all the related chemical equations.

[75] 定性分析所有的离子

Exploring Experiments

(1) Design an experiment to determine the iron content in the product using the spectrophotometric method and 1,10-phenanthroline as a chromophoric reagent[76]. You should consider all conditions such as acidity, λ_{max}, etc.

[76] 邻菲咯啉作为显色剂

(2) Design an experiment to determine the ammonium ion (NH_4^+) content in the product using titrimetric analysis.

扫一扫 视频 5-3　硫酸亚铁铵的制备及表征(讲解)
　　　　视频 5-4　硫酸亚铁铵的制备及表征(实验)

(何桂金编写)

Expt. 13　Preparation and Characterization of Copper Methionine
蛋氨酸铜的制备及表征

Objectives

(1) Understand the principles of complexometric and iodimetric titrations.

(2) Learn the difference between direct titration and back titration.

(3) Prepare a copper chelate product.

(4) Determine the copper content using complexometric titration.

(5) Determine the methionine content using iodimetric titration.

Principles

Key terms: complexometric titration (also known as complexation titration)[77], iodimetry[78], back titration[79], chelate.

Copper methionine[80], a light blue powder, is a nutritional additive[81] widely used in the feed and food industries. Due to its superior digestion and absorption, the bioavailability[82] of copper methionine is significantly higher than that of inorganic copper. It enhances immunity, reduces stress, and improves reproductive performance in animals. Additionally, copper methionine leaves minimal copper residue in animal products and causes less environmental pollution, making it a safer and more eco-friendly choice.

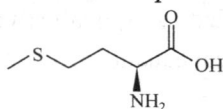

[77] 配位滴定(又称络合滴定)
[78] 直接碘量法
[79] 返滴定法
[80] 蛋氨酸铜
[81] 营养添加剂
[82] 生物利用率

Fig. 5-3　Methionine

At a pH 6-8, copper sulfate ($CuSO_4$) coordinates with methionine (Fig. 5-3) to form copper methionine precipitate. The copper content in the product can be determined by complexometric titration using a standard EDTA solution as the titrant[83].

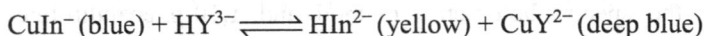

[83] 滴定剂

$$CuIn^- \text{ (blue)} + HY^{3-} \rightleftharpoons HIn^{2-} \text{ (yellow)} + CuY^{2-} \text{ (deep blue)}$$

The methionine content is determined using the back titration method and iodimetry. First, a quantitative and excess standard I_3^- solution is added to the sample. In this reaction, one molecule of I_2 reacts with one molecule of methionine, where one sulfur (S) atom combines with two iodine (I) atoms. The remaining I_3^- solution is then titrated with a standard $Na_2S_2O_3$ solution[84].

[84] 加入定量过量的 I_3^- 标准溶液(I_3^- 的作用相当于 I_2，但更稳定)。剩余的 I_3^- 标准溶液再用 $Na_2S_2O_3$ 标准溶液滴定

$$I_3^- + 2S_2O_3^{2-} = 3I^- + S_4O_6^{2-}$$

For more information about iodimetry and complexometric titration, refer to Exp. 6 and Exp. 7.

Pre-lab Questions

(1) Why is it important to control the temperature and acidity when preparing copper methionine?

(2) Why is the direct titration method unsuitable for determining the methionine content in the product?

(3) Explain the reason why a buffer[85] is always used in complexometric titration.

[85] 缓冲溶液

(4) Summarize and explain the titration errors of iodimetry.

Apparatus and Reagents

Apparatus: electronic balances (0.01 g, 0.1 mg), magnetic stirrer hot plate, vacuum pump, suction flask, Buchner funnel, burette (2×25 mL, one with red stopcock, one with blue stopcock), pipette (40 mL, 25 mL), iodine flask[86] (250 mL), volumetric flask (100 mL), Erlenmeyer flask (3×150 mL), beaker (400 mL, 250 mL), thermometer[87], stop plate[88] (black).

Reagents: $CuSO_4 \cdot 5H_2O$ (s), KIO_3 (s, primary standard), methionine (s), KI (s), H_2SO_4 (3 $mol \cdot L^{-1}$, 1 $mol \cdot L^{-1}$), HCl (2 $mol \cdot L^{-1}$), phosphate buffer[89] (pH 6.5), NaOH (5 $mol \cdot L^{-1}$), $NH_3 \cdot H_2O$ (1 : 5 or 1+5), NH_3-NH_4Cl buffer (pH 9.2), standard I_3^- solution [0.05 $mol \cdot L^{-1}$ (35 g KI + 13 g I_2, dilute to 1 L, standardize before use), standard $Na_2S_2O_3$ solution (0.1 $mol \cdot L^{-1}$), standard EDTA solution (0.05 $mol \cdot L^{-1}$), $BaCl_2$ (0.5 $mol \cdot L^{-1}$), starch (0.5%), PAN (0.1%), ethanol.

[86] 碘量瓶

[87] 温度计

[88] 点滴板

[89] 磷酸盐缓冲液

Procedures

1. Preparation of Copper Methionine

$$250 \text{ mL beaker} \xrightarrow[\text{2.3 g}]{\text{methionine}} \xrightarrow[\text{60 mL}]{\text{hot water}} \text{stir until solids dissolve} \xrightarrow[\text{1.9 g}]{CuSO_4 \cdot 5H_2O}$$

$$\text{keep stirring for 5 min at 60-70℃} \xrightarrow[\text{5 mol} \cdot L^{-1} \text{ NaOH}]{} \text{adjust pH to 6-8} \longrightarrow \text{keep stirring for 10 min at 60-70℃}$$

$$\longrightarrow \text{suction filtration and wash until } SO_4^{2-} \text{ free} \xrightarrow[\text{10 mL}]{\text{ethanol}} \text{wash twice} \longrightarrow \text{record the mass after drying}$$

Note:

(1) Control the reaction acidity. Slowly adjust the pH when it is around 4. If the pH exceeds 8, use 3 $mol \cdot L^{-1}$ H_2SO_4 to bring it down[90].

(2) Control the reaction temperature. Methionine decomposes at high temperatures, and low temperatures are unsuitable for the coordination reaction[91].

[90] 若 pH>8，用 3 $mol \cdot L^{-1}$ 硫酸反调

[91] 温度高时蛋氨酸易分解，温度低时不易配位

2. Standardization of 0.1 $mol \cdot L^{-1}$ $Na_2S_2O_3$ Solution

For more details, refer to Expt. 6.

3. Preparation of 0.02 $mol \cdot L^{-1}$ Standard EDTA Solution[92]

Accurately pipette 40.00 mL of 0.05 $mol \cdot L^{-1}$ standard EDTA solution into a 100 mL volumetric flask, then dilute to the calibration mark. This gives a 0.02 $mol \cdot L^{-1}$ standard EDTA solution.

[92] 提供的 EDTA 浓度为 0.05 $mol \cdot L^{-1}$，需先准确稀释

4. Determination of the Complex Ratio of Copper to Methionine

I. Determination of Copper Content in the Product

$$\begin{array}{l} \text{150 mL Erlenmeyer flask} \xrightarrow[\text{0.10-0.12 g}]{\text{product}} \xrightarrow[\text{30 mL}]{H_2O} \xrightarrow[\text{2 mL}]{\text{2 mol} \cdot L^{-1} \text{ HCl}} \text{heat gently until solids dissolve completely} \xrightarrow[]{\text{(1+5) } NH_3 \cdot H_2O} \\ \text{add dropwise until solution is cloudy} \xrightarrow[\text{10 mL}]{\text{buffer (pH 9.2)}} \text{heat to 80℃} \xrightarrow[\text{2-5 drops}]{\text{PAN}} \text{solution turns blue} \longrightarrow \\ \text{titrate with EDTA} \longrightarrow \text{solution turns yellowish green} \longrightarrow \begin{array}{l}\text{record the volume} \\ \text{then repeat twice}\end{array} \end{array}$$

Note:

(1) Maintain the solution at around 80°C to obtain a clear endpoint signal[93].

(2) Reheat the solution if it's difficult to detect the endpoint.

(3) Do not heat the solution to boiling[94], as this can cause methionine to decompose.

[93] 全程保持 80℃左右使终点敏锐

[94] 不能加热至沸腾

Ⅱ. Determination of Methionine Content in the Product

250 mL iodine flask	$\xrightarrow{\text{product}}$ 0.10-0.12 g	$\xrightarrow{\text{H}_2\text{O}}$ 10 mL	$\xrightarrow{2\ \text{mol}\cdot\text{L}^{-1}\ \text{HCl}}$ 3 mL	heat gently until solids dissolve completely	$\xrightarrow{\text{buffer (pH 6.5)}}$ 40 mL
mix well and cool to RT	$\xrightarrow{\text{I}_3^-\ \text{sln.}}$ 25.00 mL	cap the stopper and gently shake for 2 min	$\rightarrow$	keep in the dark for 30 min	$\xrightarrow{\text{H}_2\text{O}}$ 50 mL
$\xrightarrow{\text{Na}_2\text{S}_2\text{O}_3}$	titrate to pale yellowish green	$\xrightarrow{\text{starch}}$ 2 mL	$\xrightarrow{\text{Na}_2\text{S}_2\text{O}_3}$ titrate to azure	$\rightarrow$	record the volume then repeat twice

Note:

(1) After removing the iodine flask from the dark, immediately pour 50 mL H$_2$O to dilute the solution and prevent the volatilization of I$_2$[95].

(2) Titrate quickly and swirl gently at the start[96] to minimize the volatilization of I$_2$.

(3) Add starch near the endpoint to prevent the absorption of I$_2$.

(4) Titrate slowly and swirl vigorously near the endpoint to release the absorbed I$_2$[97].

(5) Add water as little as possible during the titration[98] to prevent I$^-$ from being oxidized by O$_2$.

[95] 立即加水稀释以减少 I$_2$ 的挥发

[96] 滴定初期快滴慢摇

[97] 临近终点慢滴快摇，将被淀粉吸附的 I$_2$ 释放出来

[98] 滴定过程中尽可能少补水

Data Treatment and Analysis

(1) Calculate the accurate concentration of the standard Na$_2$S$_2$O$_3$ solution. Calculate and analyze the relative average deviation.

(2) Calculate the percent yield, copper content, and methionine content.

(3) Estimate the ratio of copper to methionine in the product, then draw the structure of copper methionine[99].

[99] 推断产物中铜和蛋氨酸的配位比，再画出蛋氨酸铜的结构

Safety and Waste Disposal

(1) Wear a lab coat and safety goggles at all times in the laboratory. Wear gloves when handling hot glassware. If any chemicals splash into the eyes or onto the skin, wash with water immediately.

(2) Some acids or bases, such as 5 mol·L^{-1} NaOH, are corrosive and should be handled with care.

(3) Dispose of leftover products and solutions in their designated containers. Never pour oxidant or reductant waste solutions down the sink,

react them with each other first. Similarly, never pour acidic or alkaline waste solutions down the sink, neutralize them first.

Post-lab Questions

How can you improve the percent yield and purity of the product?

Exploring Experiment

(1) Draw an experiment flow chart to determine the copper content in the product using other methods.

(2) Draw an experiment flow chart to determine the methionine content in the product using other methods.

扫一扫　视频 5-5　蛋氨酸铜的制备及表征(讲解)
　　　　视频 5-6　蛋氨酸铜的制备及表征(实验)

(曾秀琼编写)

Expt. 14　Preparation and Photosensitivity Test of Potassium Trioxalatoferrate(III) Trihydrate
三水合三草酸合铁(III)酸钾的制备及光敏性测试

Objectives

(1) Understand the concept of coordination compounds[100].　　　[100] 配位化合物

(2) Prepare a particularly striking coordination compound.

(3) Test the photosensitivity[101] of the product using the cyanotype technic[102].　　　[101] 光敏性
　　　[102] 蓝晒技术

(4) Qualitatively analyze[103] the prepared compound.　　　[103] 定性分析

Principles

Key terms: coordination compound, photosensitivity, cyanotype.

Potassium trioxalatoferrate (III) trihydrate ($K_3[Fe(C_2O_4)_3 \cdot 3H_2O]$), also known as potassium iron (III) oxalate, is a magnetic organometallic coordination complex[104]. It is an emerald green monoclinic crystal with high polarity[105], dissolves easily in water but much more difficultly in ethanol. It exhibits strong photosensitivity and can also serve as the catalyst for various chemical reactions.

[104] 磁性的有机金属配合物
[105] 强极性的翠绿色单斜晶体

$Fe(NH_4)_2(SO_4)_2 \cdot 6H_2O$ or $FeSO_4 \cdot 7H_2O$ is used to prepare this product. This preparation procedure involves three chemical reactions. The first reaction is the formation of ferric (II) oxalate precipitate ($FeC_2O_4\downarrow$), the second reaction is the oxidation of FeC_2O_4 by hydrogen peroxide (H_2O_2), the third reaction leads to the formation of $K_3[Fe(C_2O_4)_3]\cdot 3H_2O$.

$$Fe^{2+} + C_2O_4^{2-} + 2H_2O = FeC_2O_4\cdot 2H_2O\downarrow \text{ (yellow)}$$

$$6FeC_2O_4\cdot 2H_2O + 3H_2O_2 + 6K_2C_2O_4 = 4K_3[Fe(C_2O_4)_3]\cdot 3H_2O\downarrow$$
$$\text{(emerald green)} + 2Fe(OH)_3\downarrow \text{ (reddish brown)}$$

$$2Fe(OH)_3 + 3K_2C_2O_4 + 3H_2C_2O_4 = 2K_3[Fe(C_2O_4)_3]\cdot 3H_2O\downarrow$$
$$\text{(emerald green)}$$

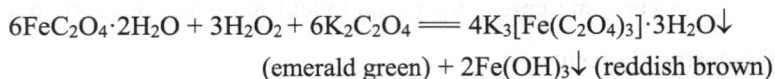

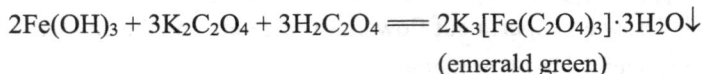

Overall reaction：

$$2Fe^{2+} + 6K^+ + 5C_2O_4^{2-} + H_2C_2O_4 + H_2O_2 + 4H_2O = 2K_3[Fe(C_2O_4)_3]\cdot 3H_2O\downarrow$$

$K_3[Fe(C_2O_4)_3]$ is highly photosensitive. When a green solution of $K_3[Fe(C_2O_4)_3]$ is exposed to sunlight for a few hours, it turns yellow due to the formation of FeC_2O_4. The complex reacts with photons[106], leading to the formation of FeC_2O_4 and CO_2. During this process, iron is reduced[107] from the +3 oxidation state[108] to the +2 oxidation state, a reaction known as photoreduction[109].

$$2[Fe(C_2O_4)_3]^{3-} \xrightarrow{hv} 2Fe^{2+} + 5C_2O_4^{2-} + 2CO_2\uparrow$$

By immersing a piece of filter paper[110] into a mixture of potassium ferricyanide ($K_3[Fe(CN)_6]$) and potassium trioxalatoferrate ($K_3[Fe(C_2O_4)_3]$), a cyanotype paper[111] can be created. When this paper is exposed to light[112], the Fe(III) salt is reduced to Fe(II) salt, which reacts with $K_3[Fe(CN)_6]$ to form Prussian blue[113], producing a permanent cyanotype photo in blue and white.

[106] 光子

[107] 被还原
[108] 氧化态
[109] 光致还原

[110] 滤纸

[111] 蓝晒纸
[112] 曝光
[113] 普鲁士蓝

Pre-lab Questions

(1) Why should all FeC_2O_4 be oxidized? How can we judge if all the FeC_2O_4 is oxidized? Give a brief description.

(2) What would happen if too much or too little H_2O_2 was added when preparing the product? Give a brief description.

(3) Should $H_2C_2O_4$ or $K_2C_2O_4$ be added when adjusting the acidity to pH 4? Give a brief explanation.

Apparatus and Reagents

Apparatus: analytical balance, hot plate, vacuum pump, suction flask, Buchner funnel, graduated cylinder, beaker, spot plate (black), boiling chips, filter paper, film, lamp, thermometer, brush.

Reagents: $(NH_4)_2Fe(SO_4)_2\cdot 6H_2O$ (s), $FeSO_4\cdot 7H_2O$ (s), H_2O_2 (10%), H_2SO_4 (3 mol·L^{-1}), $K_2C_2O_4$ (saturated solution[114]), $H_2C_2O_4$ (saturated solution), $K_3[Fe(C_2O_4)_3]$ (0.2 mol·L^{-1}), $K_3[Fe(CN)_6]$ (0.2 mol·L^{-1}), ethanol.

[114] 饱和溶液

Procedures

1. Preparation of K₃[Fe(C₂O₄)₃]·3H₂O

Ⅰ. Formation of FeC₂O₄

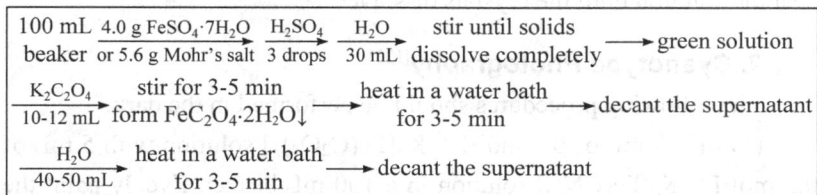

$$
\boxed{
\begin{array}{l}
\underset{\text{beaker}}{\xrightarrow{\text{100 mL}}}
\underset{\text{or 5.6 g Mohr's salt}}{\xrightarrow{\text{4.0 g FeSO}_4\cdot\text{7H}_2\text{O}}}
\underset{\text{3 drops}}{\xrightarrow{\text{H}_2\text{SO}_4}}
\underset{\text{30 mL}}{\xrightarrow{\text{H}_2\text{O}}}
\begin{array}{c}\text{stir until solids}\\ \text{dissolve completely}\end{array}
\longrightarrow \text{green solution}\\[4pt]
\underset{\text{10-12 mL}}{\xrightarrow{\text{K}_2\text{C}_2\text{O}_4}}
\begin{array}{c}\text{stir for 3-5 min}\\ \text{form FeC}_2\text{O}_4\cdot2\text{H}_2\text{O}\downarrow\end{array}
\longrightarrow
\begin{array}{c}\text{heat in a water bath}\\ \text{for 3-5 min}\end{array}
\longrightarrow \text{decant the supernatant}\\[4pt]
\underset{\text{40-50 mL}}{\xrightarrow{\text{H}_2\text{O}}}
\begin{array}{c}\text{heat in a water bath}\\ \text{for 3-5 min}\end{array}
\longrightarrow \text{decant the supernatant}
\end{array}
}
$$

Note:

(1) Ensure the temperature of the water bath is below 50°C.

(2) Use a thermometer to monitor the temperature[115].

(3) Wash the lemon-yellow precipitation twice, decanting the supernatant[116] each time.

[115] 用温度计监测温度

[116] 倾析法弃去上清液

Ⅱ. Oxidation of FeC₂O₄

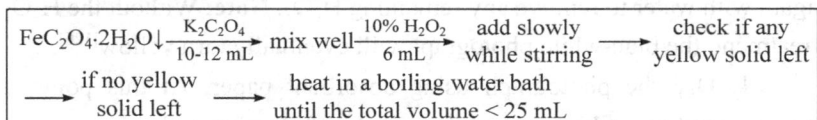

$$
\boxed{
\begin{array}{l}
\text{FeC}_2\text{O}_4\cdot2\text{H}_2\text{O}\downarrow
\underset{\text{10-12 mL}}{\xrightarrow{\text{K}_2\text{C}_2\text{O}_4}}
\text{mix well}
\underset{\text{6 mL}}{\xrightarrow{10\%\ \text{H}_2\text{O}_2}}
\begin{array}{c}\text{add slowly}\\ \text{while stirring}\end{array}
\longrightarrow
\begin{array}{c}\text{check if any}\\ \text{yellow solid left}\end{array}\\[4pt]
\longrightarrow
\begin{array}{c}\text{if no yellow}\\ \text{solid left}\end{array}
\longrightarrow
\begin{array}{c}\text{heat in a boiling water bath}\\ \text{until the total volume} < 25\text{ mL}\end{array}
\end{array}
}
$$

Note:

(1) Carefully control the amount of K₂C₂O₄ solution, as excess K₂C₂O₄ solution can react with H₂O₂.

(2) Do not add too much H₂O₂. Stir thoroughly to ensure complete oxidation of FeC₂O₄. At the end, no lemon-yellow solid should be visible at the bottom of the beaker[117].

(3) When heating the solution, use a boiling water bath[118] for safety. Ensure that the final volume in this step is less than 25 mL.

[117] 烧杯底部看不到柠檬黄色固体

[118] 沸水浴

Ⅲ. Formation of K₃[Fe(C₂O₄)₃]·3H₂O

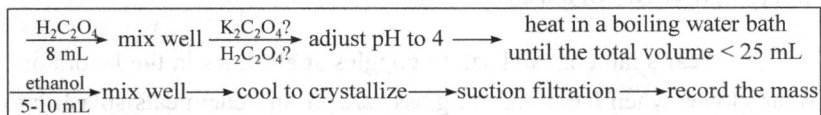

$$
\boxed{
\begin{array}{l}
\underset{\text{8 mL}}{\xrightarrow{\text{H}_2\text{C}_2\text{O}_4}}
\text{mix well}
\underset{\text{H}_2\text{C}_2\text{O}_4\text{?}}{\xrightarrow{\text{K}_2\text{C}_2\text{O}_4\text{?}}}
\text{adjust pH to 4}
\longrightarrow
\begin{array}{c}\text{heat in a boiling water bath}\\ \text{until the total volume} < 25\text{ mL}\end{array}\\[4pt]
\underset{\text{5-10 mL}}{\xrightarrow{\text{ethanol}}}
\text{mix well}
\longrightarrow \text{cool to crystallize}
\longrightarrow \text{suction filtration}
\longrightarrow \text{record the mass}
\end{array}
}
$$

Note:

(1) If heating the beaker directly on a hot plate, first add 3-4 boiling chips. Never add boiling chips to a hot solution, as this can cause the solution to be spill over[119].

(2) Carefully control the amount of H₂C₂O₄ or K₂C₂O₄ when adjusting the acidity to pH 4, as excess amounts can lead to the formation of white crystals.

[119] 不能往热溶液中加入沸石，会导致暴沸

[120] 澄清透明翠绿色溶液

[121] 微热溶液至晶体刚
好溶解

[122] 蓝晒照片

[123] 涂刷均匀

[124] 黑白负片(反色后的
胶片)

[125] 洗去残留感光剂

[126] 用 H_2O_2 浸泡可使
二价铁氧化，否则蓝白照
片会变黄

[127] 对产品进行定性分析

(3) Before crystallization, ensure the solution is clear, transparent, and emerald green[120], with no lemon-yellow or reddish-brown solid substances.

(4) If small crystals form immediately after adding ethanol, gently heat the solution until the crystals dissolve[121].

2. Cyanotype Photography[122]

The following procedures should be performed in the dark.

(1) Mix 5 mL of 0.2 mol·L^{-1} $K_3[Fe(C_2O_4)_3]$ solution with 5 mL of 0.2 mol·L^{-1} $K_3[Fe(CN)_6]$ solution in a 100 mL beaker. Evenly paint the mixture[123] onto a piece of filter paper.

(2) Place a black and white negative film[124] upside down onto the filter paper. Secure them set up with clamps, place them under the lamp, and expose to light for about 10 minutes.

(3) Rinse the photograph with tap water to remove any remaining sensitizer[125], then immerse it in a 0.3% H_2O_2 solution for a while. Rinse again with water to remove any remaining H_2O_2. **Note:** Without the H_2O_2 treatment, the blue-white photograph will eventually turn yellow[126].

(4) Dry the photograph using absorbent paper. At this point, a cyanotype photo in blue and white is successfully prepared.

3. Qualitatively Analyze the Product

Design the experimental procedures needed to perform a qualitative analysis of the product[127]. Complete this assignment before coming to the lab.

Data Treatment and Analysis

Calculate both the theoretical mass and percent yield of the product.

Safety and Waste Disposal

(1) Wear a lab coat and safety goggles at all times in the laboratory. Wear gloves when handling hot glassware. If any chemicals splash into the eyes or onto the skin, wash with water immediately.

(2) Some solutions, such as $K_3[Fe(CN)_6]$, $K_2C_2O_4$, H_2O_2, are hazardous to the human body, handle them with care.

(3) Never pour $K_3[Fe(CN)_6]$ and $K_2C_2O_4$ waste solutions down the sink. Dispose of these solutions into the designated waste containers.

Post-lab Questions

How can you obtain a large and beautiful crystal of $K_3[Fe(C_2O_4)_3]\cdot 3H_2O$?

Exploring Experiments

Design proposals to complete the following tasks.

(1) Prepare a large and beautiful $K_3[Fe(C_2O_4)_3]\cdot 3H_2O$ crystal.

(2) Test other characteristics of the product.

扫一扫　视频 5-7　三水合三草酸合铁(Ⅲ)酸钾的制备及光敏性测试(讲解)
　　　　视频 5-8　三水合三草酸合铁(Ⅲ)酸钾的制备及光敏性测试(实验)

(曾秀琼编写)

Expt. 15　Determination of the Charge Number of Trioxalatoferrate(Ⅲ) Complex Ion
三草酸合铁(Ⅲ)配离子电荷数的测定

Objectives

(1) Understand the principle and application of the 3R rule.

(2) Understand the principle of precipitation titrations[128].

(3) Learn the theory behind the Mohr method[129].

(4) Understand the principle and operation of ion exchange.

(5) Determine the charge number of a trioxalatoferrate(Ⅲ) complex ion[130].

[128] 沉淀滴定

[129] 莫尔法

[130] 三草酸合铁(Ⅲ)配离子电荷数

Principles

Key terms: anion-exchange[131], precipitation titration, argentimetry[132], Mohr method.

The complex anions in a $K_3[Fe(C_2O_4)_3]\cdot 3H_2O$ solution can be exchanged with anion exchange resins. During this process, the complex anions are absorbed by the resins, while chloride anions (Cl^-) from the resins are released into the solution[133].

$$zRN^+Cl^- + X^{z-} \rightleftharpoons (RN^+)_zX^{z-} + zCl^-$$

The concentration of Cl^- can be determined using Mohr method, one kind of argentimetry. For more details about precipitation titration and argentimetry, refer to Expt. 8.

In this experiment, silver nitrate ($AgNO_3$) is the titrant[134], and potassium chromate (K_2CrO_4) serves as the indicator. The endpoint of the titration is reached when all the chloride anions (Cl^-) have been precipitated.

[131] 阴离子交换

[132] 银量法

[133] 配阴离子与树脂上Cl⁻交换，被吸附在树脂上，Cl⁻被取代后进入溶液中

[134] 滴定剂

Afterward, the excess Ag^+ cations react with chromate ions (CrO_4^{2-}) to form a red-brown precipitate of silver chromate (Ag_2CrO_4)[135].

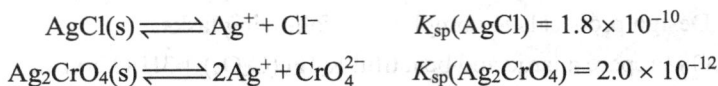

$$AgCl(s) \rightleftharpoons Ag^+ + Cl^- \qquad K_{sp}(AgCl) = 1.8 \times 10^{-10}$$

$$Ag_2CrO_4(s) \rightleftharpoons 2Ag^+ + CrO_4^{2-} \qquad K_{sp}(Ag_2CrO_4) = 2.0 \times 10^{-12}$$

Finally, the charge number of the complex anion can be calculated using the following equation.

$$z = \frac{n_{Cl^-}}{n_{complex}}$$

Pre-lab Questions

(1) How does an indicator for precipitation titration function? Describe briefly.

(2) List the requirements of Mohr method. Briefly explain each requirement.

(3) List the key steps to obtain a good measurement result.

Apparatus and Reagents

Apparatus: analytical balance, volumetric flask (brown, 50 mL), burette (brown, 10 mL), ion exchange column, Erlenmeyer flask (3×150 mL), graduated cylinder, beaker, spot plate (black).

Reagents: $K_3[Fe(C_2O_4)_3]\cdot 3H_2O$ (s), $AgNO_3$ (s, 0.1 $mol\cdot L^{-1}$), K_2CrO_4 (0.1 $mol\cdot L^{-1}$), Cl^--anion exchange resin (soak in a 1 $mol\cdot L^{-1}$ NaCl solution for 48 h before use, rinse with deionized water until Cl^- free[136]).

Procedures

1. Setting Up the Apparatus

exchange column	$\xrightarrow[\text{8-10 cm high}]{\text{ion exchange resin}}$	$\xrightarrow{H_2O}$ wash the resin until Cl^- free	$\longrightarrow$ set a 50 mL volumetric flask

Note:

(1) Wash the resin several times until it is free of Cl^-.

(2) Do not allow the resin to dry or form air bubbles.

2. Exchanging and Collection

100 mL beaker	$\xrightarrow[\text{0.20-0.24 g}]{\text{product}}$	$\xrightarrow{H_2O}$ 8 mL	stir until solids dissolve completely	$\longrightarrow$ transfer to the column	$\longrightarrow$ adjust fluid speed $\longrightarrow$
collect in 50 mL volumetric flask	$\xleftarrow{H_2O}$	rinse the beaker and glass rod 2-3 times	$\xleftarrow{H_2O}$	rinse the resin until Cl^- free	dilute to the mark

Note:

(1) Do not use too much water to dissolve the product.

(2) First transfer all of the product solution, then rinse with deionized water. Use a glass rod to carefully transfer the solution or water each time, ensuring not to splash the resin[137].

[137] 小心转移,不能将树脂溅起

(3) Control the exchange speed[138]; if it's too fast, the complex anion may drain without sufficient exchange.

[138] 交换速率

(4) Do not allow the resin to dry or form air bubbles.

3. Preparation of 0.04 mol·L^{-1} Standard AgNO$_3$ Solution

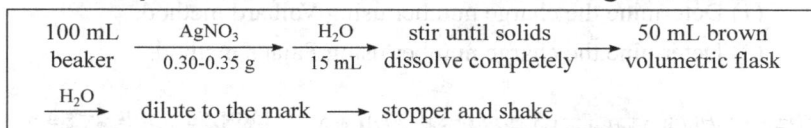

Note:

(1) Due to the semimicro-quantitative analysis[139], use the direct method to prepare this standard AgNO$_3$ solution. If using the indirect method, refer to Expt. 8.

[139] 半微量定量分析

(2) All glassware used in this step must be thoroughly cleaned and rinsed several times with deionized water. Otherwise, the AgNO$_3$ solution will be muddy[140].

[140] 否则，AgNO$_3$ 溶液是浑浊的

4. Titration

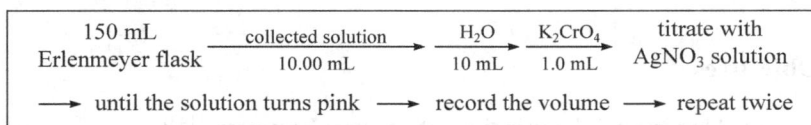

Note:

(1) Use an automatic pipette to add[141] 1.0 mL K$_2$CrO$_4$.

[141] 用移液枪加入

(2) The color at the endpoint should be pink, not brick red.

(3) Titrate slowly as you approach the endpoint.

Data Treatment and Analysis

Calculate the charge number of the complex ion and analyze the measurement error.

Safety and Waste Disposal

(1) Wear a lab coat and safety goggles at all times in the laboratory. Wear gloves when handling hot glassware. If any chemicals splash into the eyes or onto the skin, wash with water immediately.

(2) Some solutions, like K$_2$CrO$_4$, AgNO$_3$, are hazardous to the human body, handle with care. The AgNO$_3$ solution can easily stain skin, clothing, and other surfaces[142].

[142] 易沾污皮肤、衣物和其他表面

(3) AgNO$_3$ is an expensive chemical, so pour the excess AgNO$_3$ solution into the designated container for reuse.

Post-lab Questions

Why should K_2CrO_4 solution be added precisely? Give a brief explanation.

Exploring Experiments

Design proposals to complete the following tasks.
(1) Determine the charge number using Volhard method.
(2) Determine the charge number using Fajans method.

扫一扫　视频 5-9　三草酸合铁(Ⅲ)配离子电荷数的测定(讲解)
　　　　视频 5-10　三草酸合铁(Ⅲ)配离子电荷数的测定(实验)

(曾秀琼编写)

Expt. 16　Spectrophotometric Determination of Iron Content
分光光度法测定铁含量

Objectives

[143] 分光光度法

[144] 条件优化

[145] 分光光度计

[146] 配合物组成

(1) Understand the principle of spectrophotometry[143].
(2) Optimize the conditions[144] for spectrophotometric determination.
(3) Learn how to operate a spectrophotometer[145].
(4) Determine the iron content using the spectrophotometric method.
(5) Study the composition of the iron-Phen complex[146].

Principles

[147] 标准曲线、吸收曲线(光谱)、参比溶液

[148] 血红蛋白

[149] 与邻菲咯啉(邻二氮菲)形成稳定的配合物

Key terms: spectrophotometry, spectrophotometer, Lambert-Beer law, standard curve, absorbance curve (spectrum), reference solution[147].

Fe^{2+} in hemoglobin[148] is crucial for the transport and storage of O_2 in the blood. Therefore, it is essential that iron remains in its Fe^{2+} (ferrous), rather than being oxidized to Fe^{3+} (ferric) in the human body. This experiment introduces a method for determining the content of Fe^{2+} and Fe^{3+} in solutions. Ferrous irons (Fe^{2+}) form stable complexes with phenanthroline (Phen)[149] within the pH range of 3-9. These complexes, with the chemical formula $[Fe(C_{12}H_8N_2)_3]^{2+}$, exhibit a reddish-orange color with $\lambda_{max} = 510$ nm. Therefore, the iron content can be determined using spectrophotometric methods.

$$Fe^{2+} + 3 \quad \longrightarrow \quad \left[\left(\bigcirc \bigcirc \right)_3 Fe \right]^{2+}$$

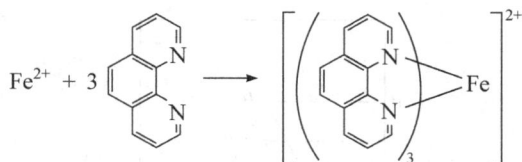

When determining the total iron content, hydroxylamine hydrochloride ($NH_2OH \cdot HCl$) is used to reduce the ferric iron (Fe^{3+}) to ferrous iron (Fe^{2+})[150].

$$2Fe^{3+} + 2NH_2OH \cdot HCl \Longrightarrow 2Fe^{2+} + N_2 + 2H_2O + 4H^+ + 2Cl^-$$

The selectivity of this method is excellent. The detection selectivity[151] for Fe is 40 times greater than for Sn, Al, Ca, Mg, Zn, SiO_3^{2-}, 20 times greater than for Cr, Mn, V, PO_4^{3-}, and 5 times greater than for Co, Cu.

When performing spectrophotometric measurement, it is crucial to consider several factors: the absorbance curve, the concentration of the chromophoric reagent[152], the stability of the colored substance, the acidity of the solution, the standard curve, and more. In this experiment, some of these factors are initially explored and then confirmed to establish optimal conditions[153].

Spectrophotometry can be used to determine the stoichiometry of metal-ligand complexes[154]. There are two common methods for this purpose.

(1) **Mole ratio method**[155]: In this method, the amount of one reactant, typically the moles of metal, is kept constant, while the amount of the other reactant is varied. Absorbance is monitored at a wavelength where the metal-ligand complex absorbs, and a plot is generated[156]. Fig. 5-4 shows a typical result for the formation of a 1 : 1 complex.

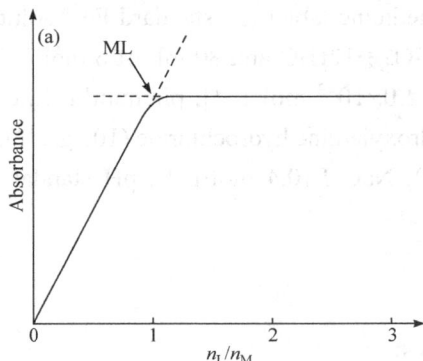

Fig. 5-4   Mole ratio method

(2) **Continuous variation method (Job's method)**[157]: Developed by Paul Job in 1928, this method keeps the total concentration of two reactants constant while varying their proportions. Fig. 5-5 shows typical results for the formation of a 1 : 1 complex.

[150] 盐酸羟胺用于把Fe^{3+}还原成Fe^{2+}

[151] 检测灵敏度

[152] 显色剂浓度

[153] 先探究再确定最佳实验条件

[154] 金属与配体的化学计量比

[155] 摩尔比法:保持一种反应物(通常是金属)的物质的量不变

[156] 作图

[157] 连续变化法(等摩尔系列法):两种反应物浓度之和保持不变,改变其中一种反应物浓度

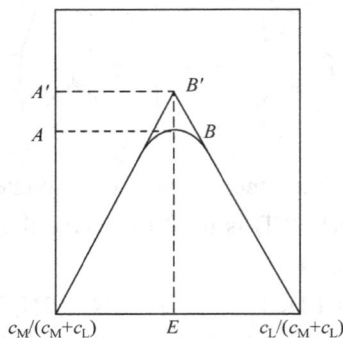

Fig. 5-5　Continuous variation method

Pre-lab Questions

(1) Briefly summarize the main structure and principles of a spectrophotometer[158].

[158] 简单描述分光光度计的主要结构和原理

(2) What are the purposes of measuring an absorbance curve?

(3) Why should the absorbance be measured at λ_{max}?

[159] 参比溶液

(4) Why and how should a reference solution[159] be selected?

[160] 解离常数

(5) Using the continuous variation method, you can obtain the dissociation constant[160] of this iron-complex. Give a brief explanation.

Apparatus and Reagents

[161] 超声波振荡器
[162] 移液枪
[163] 比色管
[164] 分度吸量管
[165] 单标线吸量管
[166] 研钵

Apparatus: spectrophotometer, pH meter, ultrasonic oscillator[161], automatic pipette[162], colorimetric tube[163], graduated pipette[164] (1 mL, 2 mL, 5 mL), volumetric pipette[165] (10 mL), graduated cylinder (10 mL, 50 mL), glass funnel with long neck, mortar[166], cuvette (1 cm).

Reagents: medicine tablet (s), standard Fe^{3+} solution [100 mg·L^{-1}: 0.8634 g $NH_4Fe(SO_4)_2 \cdot 12H_2O$ and 80 mL of 6 mol·L^{-1} HCl to 1000 mL of total volume; 2.0×10^{-3} mol·L^{-1}], phenanthroline (1.5 g·L^{-1}, 2.0×10^{-3} mol·L^{-1}), hydroxylamine hydrochloride (100 g·L^{-1}), HCl (2 mol·L^{-1}), NaAc (1 mol·L^{-1}), NaOH (0.4 mol·L^{-1}), pH standard buffer solutions (4.00, 6.86, 9.18).

Procedures

General notes:

(1) Add the solutions sequentially from left to right as listed in the table. Shake to mix well after adding hydroxylamine hydrochloride ($NH_2OH \cdot HCl$)[167].

[167] 表格中试剂从左到右依次加入。加完盐酸羟胺后需振摇(使 Fe^{3+} 还原完全)

(2) All cuvettes used in this experiment have a light path length of 1 cm.

(3) Handle the cuvettes with care, as they are fragile.

1. Exploring Optical Determination Conditions

Ⅰ. Determination of the Absorbance Curve

Fill three cuvettes[168] with the solutions $3^\#$, $5^\#$ and $7^\#$ from Table 5-7. Measure the absorbance of each solution from 460 nm to 540 nm and record the data in Table 5-3. Plot an A-λ curve (absorbance curve)[169] to determine the λ_{max} for the following measurements.

[168] 比色皿

[169] 绘制吸收曲线

Table 5-3   Determination of the absorbance curve

λ/nm	460	470	480	490	500	506	508	510	520	530	540
A											

Note:

(1) Whenever the wavelength is changed, recalibrate the spectrophotometer using the reference solution[170].

(2) If the spectrophotometer has an automatic scan function[171], use it to determine the λ_{max} directly.

[170] 用参比溶液重新校准分光光度计

[171] 自动扫描功能

Ⅱ. Determination of the Stability of the Colored Complex

Measure the absorbance of solutions $5^\#$ at λ_{max} over time, recording the absorbance at 0, 1, 3, 5, 10, 20, 35, and 60 minutes. Record the data in Table 5-4. Plot an A-t curve (stability curve) to analyze the stability of the colored complex and determine the optimal reaction time[172].

[172] 分析有色配合物的稳定性，确定最佳反应时间

Table 5-4   Stability of the colored complex

t /min	0	1	3	5	10	20	35	60
A								

Ⅲ. Effect of the Concentration of Chromophoric Reagent[173]

[173] 显色剂浓度

Table 5-5   Effect of the concentration of chromophoric reagent

V(Phen)/mL	0.30	0.50	1.00	1.50	2.00	3.00	5.00
A							

Note:

(1) Accurately pipette standard Fe^{3+} and Phen solutions.

(2) Check if the cuvettes are matched[174]. Fill each cuvette 1/2 to 1/3 full with water. Using one as a reference, measure the absorbance of others at the same wavelength. If the absorbance difference is within ±0.005, the cuvettes are considered matched[175].

[174] 检测比色皿是否配套

[175] 若吸光度差值在 ±0.005 内，则比色皿视为配套

IV. Effect of the pH of the Solution

Step 1. Preparation of the stock solution.

$$
\begin{array}{c}
\text{100 mL} \xrightarrow[\text{10 mL}]{\text{100 mg·L}^{-1}\text{ Fe}^{3+}} \xrightarrow[\text{10 mL}]{\text{NH}_2\text{OH·HCl}} \xrightarrow[\text{5 mL}]{\text{2 mol·L}^{-1}\text{ HCl}} \begin{array}{l}\text{shake to mix well}\\ \text{keep still for 2 min}\end{array}\\
\text{volumetric flask}\\[4pt]
\xrightarrow[\text{20 mL}]{\text{1.5 g·L}^{-1}\text{ Phen}} \text{dilute to the mark}
\end{array}
$$

[176] 此步可用量筒量取
所有溶液(因为只是配制
储备液)

Note: Use graduated cylinders to measure all the above solutions[176].

Step 2. Effect of pH.

$$
\begin{array}{c}
\text{8×50 mL} \xrightarrow[\text{10.00 mL}]{\text{above solution}} \xrightarrow[\substack{\text{2.00, 3.00, 4.50, 6.30, 6.70, 8.00 mL, see Table 5-6)}}]{0.40\text{ mol·L}^{-1}\text{NaOH (0.00, 1.00,}}\\
\text{colorimetric tube}\\[4pt]
\text{dilute to the mark} \longrightarrow \text{measure pH} \longrightarrow \substack{\text{measure}\\ A\text{ at }\lambda_{max}} \longrightarrow \text{plot an }A\text{-pH curve}\\[4pt]
\longrightarrow \text{find the optimal pH range}
\end{array}
$$

Table 5-6   Effect of pH

V(NaOH)/ mL	0.00	1.00	2.00	3.00	4.50	6.30	6.70	8.00
pH								
A								

Note:

(1) Accurately pipette NaOH solutions.

(2) If the acidity or alkalinity of the solution is too low, gently shake the colorimetric tube to obtain a fast and stable reading from the pH meter[177].

[177] 当溶液酸碱性较弱
时(电极响应很慢)，轻轻摇
动溶液以加快电极平衡

2. Determination of the Iron Content in a Given Unknown Solution

I. Determination of the Standard Curve

$$
\begin{array}{c}
\text{7×50 mL} \xrightarrow[\substack{\text{0.40, 0.80, 1.20, 1.60, 2.00 mL)}}]{\text{100 mg·L}^{-1}\text{ Fe}^{3+}\text{ (0.00, 0.20,}} \xrightarrow[\text{1 mL}]{\text{NH}_2\text{OH·HCl}} \substack{\text{shake to}\\ \text{react well}} \xrightarrow[\text{5 mL}]{\text{NaAc}}\\
\text{colorimetric tube}\\[4pt]
\xrightarrow[\text{2 mL}]{\text{Phen}} \text{dilute to the mark} \longrightarrow \substack{\text{using 1}^{\#}\text{ as reference}\\ \text{measure }A\text{ at }\lambda_{max}} \longrightarrow \text{plot an }A\text{-}c_{\text{Fe}}\text{ curve}\\[4pt]
\longrightarrow \text{show the equation and }R^2\text{ of the line}
\end{array}
$$

Note:

(1) Accurately pipette standard Fe^{3+} solutions.

(2) First, check if the cuvettes are matched.

II. Determination of the Iron Content

Prepare solution $8^{\#}$ from Table 5-7, measure its absorbance, and determine the concentration using the standard curve[178].

[178] 利用标准曲线测得
其浓度

Table 5-7 Iron Determination

No.	V(standard Fe solution) /mL	V(given unknown) /mL	V(medicine solution) /mL	$V(NH_2OH \cdot HCl)$/mL	V(NaAc) /mL	V(Phen) /mL	A	$c(Fe^{2+})$ /(mg·L^{-1})
1	0.00	0.00	0.00	1.00	5.00	2.00		
2	0.20	0.00	0.00	1.00	5.00	2.00		
3	0.40	0.00	0.00	1.00	5.00	2.00		
4	0.80	0.00	0.00	1.00	5.00	2.00		
5	1.20	0.00	0.00	1.00	5.00	2.00		
6	1.60	0.00	0.00	1.00	5.00	2.00		
7	2.00	0.00	0.00	1.00	5.00	2.00		
8	0.00	5.00	0.00	1.00	5.00	2.00		
9	0.00	0.00	5.00	1.00	5.00	2.00		

3. Determination of the Iron Content in a Tablet of Medicine

Ⅰ. Preparation of the Medicine Solution[179]

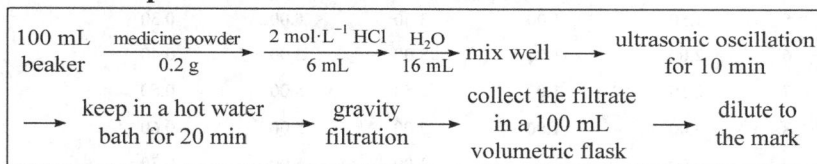

100 mL beaker →(medicine powder 0.2 g)→ 2 mol·L^{-1} HCl 6 mL →(H$_2$O 16 mL)→ mix well → ultrasonic oscillation for 10 min → keep in a hot water bath for 20 min → gravity filtration → collect the filtrate in a 100 mL volumetric flask → dilute to the mark

Note:

(1) Accurately weigh four tablets of the medicine (to the nearest 0.1 mg), and gently grind them into fine powder[180].

(2) The purpose of ultrasonic oscillation[181] is to improve dissolution.

(3) The purpose of heating is to enhance the aging and improve the filtration[182].

(4) Rinse the beaker and glass rod 2-3 times, each time transferring the rinsed solution to the filter paper.

Ⅱ. Determination of the Iron Content

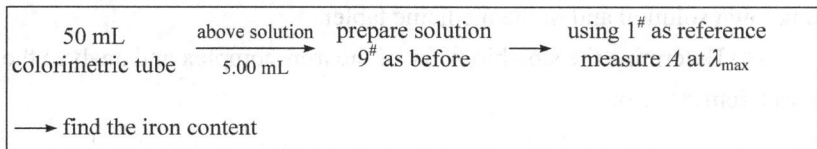

50 mL colorimetric tube →(above solution 5.00 mL)→ prepare solution 9# as before → using 1# as reference measure A at λ_{max} → find the iron content

4. Study of Composition of the Iron Complex

The composition (the stoichiometry) of this iron-complex can be studied using the mole ratio method and the continuous variation method.

Ⅰ. Mole Ratio Method

Use the data from Table 5-5 to plot an A-c_{Phen}/c_M curve and determine

[179] (操作要点)加热陈化后,需先冷却;为加快常压过滤,需先做好一段水柱;沉降后将上清液转移到漏斗中,固体留在烧杯中

[180] 研成粉末

[181] 超声波振荡

[182] 促进陈化和过滤

the stoichiometry[183] of the iron complex.

Ⅱ. Continuous Variation Method

Use 2.0×10^{-3} mol·L^{-1} standard Fe^{3+} solution and 2.0×10^{-3} mol·L^{-1} Phen solution to prepare a series solution from 1$^{\#}$ to 10$^{\#}$ in ten 50 mL colorimetric tubes, diluting to the mark. Use solution 1$^{\#}$ as the reference and measure the absorbance of solutions 2$^{\#}$ to 10$^{\#}$ at the previous determined λ_{max}. Record the data in Table 5-8, and plot a curve of A vs $c_M/(c_M + c_{Phen})$ to determine the stoichiometry of the iron complex.

Table 5-8　Determination of the Stoichiometry

No.	V(standard Fe solution) /mL	V(NH$_2$OH·HCl)/mL	V(Phen)/mL	V(NaAc) /mL	$c_M/(c_M + c_{Phen})$	A
1	0.00	1.00	5.00	5.00	0.00	
2	0.50	1.00	4.50	5.00	0.10	
3	0.75	1.00	4.25	5.00	0.15	
4	1.00	1.00	4.00	5.00	0.20	
5	1.50	1.00	3.50	5.00	0.30	
6	2.00	1.00	3.00	5.00	0.40	
7	2.50	1.00	2.50	5.00	0.50	
8	3.00	1.00	2.00	5.00	0.60	
9	4.00	1.00	1.00	5.00	0.80	
10	5.00	1.00	0.00	5.00	1.00	

Note: Pay attention to the concentrations of Fe^{3+} and Phen solutions, as they differ from those in other steps[184].

Data Treatment and Evaluations

(1) Use EXCEL or ORIGIN software to plot all the curves.

(2) Based on the data and curves, summarize the optimal condition for each step[185].

(3) Using the standard curve, calculate the iron content in the given unknown solution and in the medicine tablet.

(4) Determine the stoichiometry of the iron-complex and analyze the measurement error.

Safety and Waste Disposal

(1) Wear a lab coat and safety goggles at all times in the laboratory. If any chemicals splash into the eyes or onto the skin, wash with water immediately.

(2) NH$_2$OH·HCl is harmful and it must be handled with care. Wear gloves whenever needed.

(3) Do not pour the acidic or alkaline waste solutions down the sink. First neutralize them with each other.

Post-lab Questions

(1) Calculate the dissociation constant[186] of this iron complex.

(2) Calculate the crystal field splitting energy[187] of this iron complex.

[186] 解离常数

[187] 晶体场分裂能

Exploring Experiment

Besides standard curve method, standard addition method[188] can be used to determine the unknown sample. As shown in Table 5-9, add some unknown to the standard solution, then measure the absorbance.

[188] 标准加入法

Table 5-9 Determination of the iron content by standard addition method

	$V(Fe^{3+})$ /mL	V(medicine) /mL	$V(NH_2OH·HCl)$ /mL	V(NaAc) /mL	V(Phen) /mL	Conc. added /(mg·L^{-1})	A
1	0.00	0.00	1.00	5.00	2.00		
2	0.00	1.00	1.00	5.00	2.00		
3	0.20	1.00	1.00	5.00	2.00		
4	0.40	1.00	1.00	5.00	2.00		
5	0.80	1.00	1.00	5.00	2.00		
6	1.20	1.00	1.00	5.00	2.00		
7	1.60	1.00	1.00	5.00	2.00		

For the unknown (unk), $A_{unk} = abc_{unk}$

For the unknown and standard solution (std), $A = ab(c_{unk} + c_{std})$

Therefore, $c_{unk} = c_{std} /[(A/A_{unk}) -1]$

For better results, prepare a series of standard solutions (just as in a typical Beer's law analysis), each adding the same volume of the unknown. Plot a curve of concentration (conc.) added to the unknown vs. absorbance. The negative of the x-intercept gives the concentration of the unknown[189].

[189] 以加入标准溶液的浓度对吸光度作图，x 轴截距的负数即为未知液的浓度

Task:

(1) Design an experiment using the standard addition method to determine the iron content in the medicine tablet.

(2) Compare and analyze the result obtained from the standard curve method and the standard addition method.

(3) Describe briefly the advantages and disadvantages of these two methods.

扫一扫　视频 5-11　分光光度法测定条件优化(讲解)
视频 5-12　分光光度法测铁(讲解)
视频 5-13　分光光度法测铁(实验)

（曾秀琼编写）

Expt. 17 Synthesis and Characterization of Three Kinds of Cobalt-ammine Coordination Compounds
三种钴氨配合物的制备及表征

Objectives

[190] 配位化学

(1) Know the foundation of coordination chemistry[190].

(2) Synthesize three different cobalt-amine coordination compounds.

[191] 间接碘量法

(3) Determine the cobalt content in the products using iodometry[191].

[192] 用电导法分析产物的离子构型

(4) Analyze the ion configurations of the products using conductometry[192].

[193] 用分光光度法测定晶体场分裂能

(5) Measure the crystal field splitting energy of the products using spectrophotometry[193].

Principles

Key terms: coordination compound, iodometry, conductometry, crystal field splitting energy.

[194] 异构体

There are many isomers[194] of cobalt (III) coordination compounds

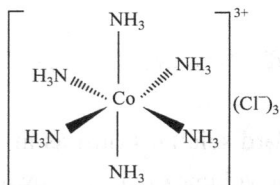

Fig. 5-6 [Co(NH₃)₆]Cl₃

[195] 维尔纳配合物的原型

due to their inner composition (three examples are shown in Table 5-10). $[Co(NH_3)_6]Cl_3$ (Fig. 5-6) is considered an archetypal "Werner complex"[195], named after Alfred Werner, the pioneer of coordination chemistry. In 1898, Swiss chemist Alfred Werner developed the foundational concepts and structures of coordination compounds and proposed the coordination theory. He was awarded the Nobel Prize in Chemistry in 1913, the first ever for Inorganic Chemistry.

Table 5-10   Three isomers of cobalt (III) coordination compounds

Compounds	Name	Color	Molar mass /(g·mol⁻¹)
$[Co(NH_3)_5Cl]Cl_2$	chloropentaaminecobalt (III) chloride	purple red	250.4
$[Co(NH_3)_6]Cl_3$	hexaamminecobalt (III) chloride	orange yellow	267.5
$[Co(NH_3)_5H_2O]Cl_3$	pentaammineaquocobalt (III) chloride	brick red	268.5

[196] 活性炭的催化和过氧化氢的氧化

Under the catalysis of activated charcoal and the oxidation of hydrogen peroxide (H_2O_2)[196], $[Co(NH_3)_6]Cl_3$ can be synthesized using cobalt (II) chloride $(CoCl_2)$ and concentrated ammonium solution $(NH_3 \cdot H_2O)$ as starting materials. Based on the principles of chemical equilibrium[197] and the solubility product[198], $[Co(NH_3)_6]Cl_3$ crystals can be precipitated from a concentrated hydrochloric acid (HCl) solution[199].

    By adjusting the reaction temperatures, $[Co(NH_3)_5H_2O]Cl_3$ and $[Co(NH_3)_5Cl]Cl_2$ can be synthesized from the same raw material without the need for activated charcoal.

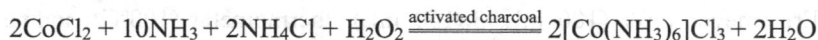

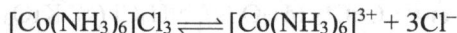

$$2CoCl_2 + 10NH_3 + 2NH_4Cl + H_2O_2 \xrightarrow{\text{activated charcoal}} 2[Co(NH_3)_6]Cl_3 + 2H_2O$$
$$2CoCl_2 + 8NH_3 + 2NH_4Cl + H_2O_2 \Longrightarrow 2[Co(NH_3)_5Cl]Cl_2 + 2H_2O$$
$$2CoCl_2 + 8NH_3 + 2NH_4Cl + H_2O_2 \Longrightarrow 2[Co(NH_3)_5H_2O]Cl_3$$
$$[Co(NH_3)_6]Cl_3 \Longrightarrow [Co(NH_3)_6]^{3+} + 3Cl^-$$

    Iodometry[200] is used to determine the cobalt content in the products. When boiled in an excess strong alkaline solution[201], cobalt(III) coordination compound decomposes[202] to forms $Co(OH)_3$.

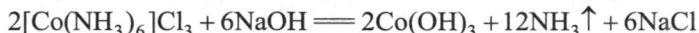

$$2[Co(NH_3)_6]Cl_3 + 6NaOH \Longrightarrow 2Co(OH)_3 + 12NH_3\uparrow + 6NaCl$$
$$2[Co(NH_3)_5Cl]Cl_2 + 6NaOH \Longrightarrow 2Co(OH)_3 + 10NH_3\uparrow + 6NaCl$$
$$2[Co(NH_3)_5H_2O]Cl_3 + 6NaOH \Longrightarrow 2Co(OH)_3\downarrow + 10NH_3\uparrow + 6NaCl + 2H_2O$$

    Due to its oxidizing properties[203], $Co(OH)_3$ can oxidize iodide (I^-) to iodine (I_2). The iodine produced can be titrated using a standard sodium thiosulfate ($Na_2S_2O_3$) solution. In this way, the cobalt (Co) content in the products is determined.

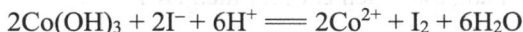

$$2Co(OH)_3 + 2I^- + 6H^+ \Longrightarrow 2Co^{2+} + I_2 + 6H_2O$$
$$2S_2O_3^{2-} + I_2 \Longrightarrow S_4O_6^{2-} + 2I^-$$

    The conductivity[204] of each cobalt (III) coordination compound can be measured using a conductivity meter[205], allowing its ion configuration to be deduced (Table 5-11).

Table 5-11 Relationship between molar conductivity[206] and ion configuration
$(1.0\times10^{-3}\ mol\cdot L^{-1}, 25°C)$

Ion configuration	Ion number	Conductivity$\Lambda/(\mu S\cdot cm^{-1})$
MA	2	120-134
MA_2 or M_2A	3	240-278
MA_3 or M_3A	4	411-451
MA_4 or M_4A	5	533-569

Pre-lab Questions

    (1) What are the functions of H_2O_2, $NH_3\cdot H_2O$ and concentrated HCl in the synthesis procedures?

    (2) What is the difference between iodometry and iodimetry[207]? What are the key points to prevent the measurement error of iodometry in this experiment?

    (3) Give the determination principle of cobalt content in the product.

---

[197] 化学平衡
[198] 溶度积
[199] 从浓盐酸中沉淀

[200] 间接碘量法
[201] 在过量强碱溶液中煮沸

[202] 分解

[203] 氧化能力

[204] 电导率
[205] 电导率仪

[206] 摩尔电导率

[207] 间接碘量法和直接碘量法

Apparatus and Reagents

[208] 分光光度计

Apparatus: analytical balance, spectrophotometer[208], oven, conductivity meter, water bath, hot plate, magnetic stirrer, vacuum pump, Buchner funnel, suction flask, volumetric flask, burette (25 mL), Erlenmeyer flask (100 mL, 3×150 mL), thermometer (100°C), iodine flask[209] (250 mL), Erlenmeyer conical flask with ground-in stopper[210] (250 mL), parafilm[211].

[209] 碘量瓶

[210] 具塞锥形瓶(没有杯状水封圈,而碘量瓶有)

[211] 封口膜

Reagents: $CoCl_2 \cdot 6H_2O$ (s), NH_4Cl (s), KIO_3 (s, primary standard), KI (s), activated charcoal (s), HCl (concentrated, 6 mol·L^{-1}), H_2SO_4 (3 mol·L^{-1}), $NH_3 \cdot H_2O$ (concentrated), NaOH (5 mol·L^{-1}), H_2O_2 (10%), $Na_2S_2O_3$ (0.1 mol·L^{-1}), starch (0.5%), ice.

Procedures

General notes:

[212] 在通风橱内操作浓盐酸和浓氨水

(1) Add concentrated HCl and $NH_3 \cdot H_2O$ in the fume hood[212].

(2) After each addition of concentrated HCl and $NH_3 \cdot H_2O$, cover the neck of the Erlenmeyer flask with a piece of parafilm[213] to prevent the release of harmful gases.

[213] 用封口膜包裹锥形瓶瓶口

(3) Do not add too much concentrated HCl.

1. Synthesis of [Co(NH₃)₆]Cl₃

100 mL Erlenmeyer flask	$\xrightarrow[4.0 \text{ g}]{NH_4Cl}$	$\xrightarrow[6.0 \text{ g}]{CoCl_2 \cdot 6H_2O}$	$\xrightarrow[8 \text{ mL}]{H_2O}$	swirl to mix well ⟶

heat in a water bath at 60°C for 10 min ⟶ cool to RT $\xrightarrow[0.4 \text{ g}]{\text{activated charcoal}}$

swirl for several minutes $\xrightarrow[10 \text{ mL}]{\text{conc. } NH_3 \cdot H_2O}$ keep in ice bath $\xrightarrow[8 \text{ mL}]{H_2O_2}$

add slowly while swirling gently ⟶ heat in a water bath at 60°C for 15 min ⟶

suction filtration ⟶ get the sediment

---

100 mL beaker $\xrightarrow[40-60 \text{ mL}]{\text{boiling HCl (3+50)}}^{\text{sediment}}$ stir quickly ⟶ suction filtration

100 mL beaker $\xrightarrow[10 \text{ mL}]{\text{conc. HCl}}^{\text{filtrate solution}}$ mix well ⟶ cool in an ice bath ⟶

suction filtration ⟶ place the product in a watch glass ⟶ dry at 90°C for 20-30 min ⟶

weigh and record the mass

Note:

[214] 小心称量活性炭,极易撒出

(1) Weigh the activated charcoal carefully, as it is easy to spill. [214]

[215] 不时振摇

(2) Maintain the Erlenmeyer flask at around 60°C, swirling it occasionally[215] to help the reaction reach completion.

(3) Prepare the boiling HCl (3+50) just before use by adding 3 droppers of concentrated HCl to 50 mL of boiling water[216].

[216] 将 3 滴管浓盐酸加入 50 mL 沸水中

(4) The required volume of boiling HCl (3+50) may vary with room temperature, less in hot weather and more in cold weather.

(5) Ensure that no yellow solids remain in the reaction solution before performing suction filtration[217].

[217] 抽滤前需确保反应溶液中无黄色固体残留

2. Synthesis of [Co(NH₃)₅Cl]Cl₂

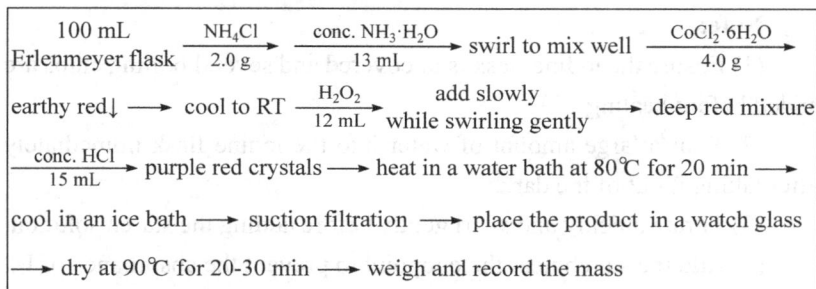

Note: Keep the Erlenmeyer flask at around 80ºC, swirling it occasionally to help the reaction reach completion[218].

[218] 保持较高温度并确保反应完全

3. Synthesis of [Co(NH₃)₅H₂O]Cl₃

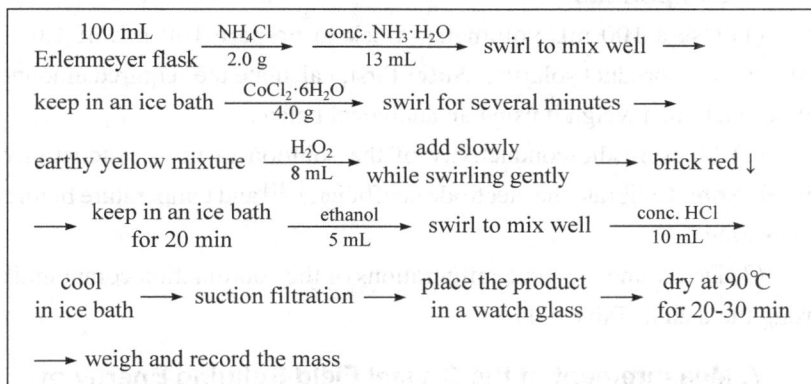

Note:

(1) Maintaining a low temperature is crucial. The Erlenmeyer flask must be kept in an ice bath at all times[219].

[219] 全程冰浴

(2) Swirl the Erlenmeyer flask occasionally to help the reaction reach completion.

4. Standardization of 0.1 mol·L⁻¹ Na₂S₂O₃ Solution

Refer to Expt. 6, using KIO₃ as a primary standard.

[220] (操作要点)加入沸石后煮沸，不能密封碘量瓶。从暗处取出后立即加入大量水稀释碘单质。加淀粉前快滴慢摇。临近终点加入淀粉，然后慢滴快摇以释放被淀粉吸附的 I_2

5. Determination of the Cobalt Content in the Product[220]

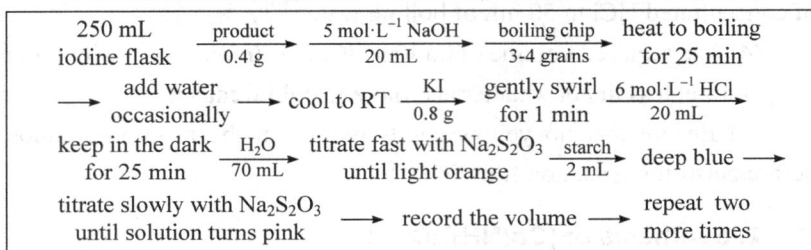

250 mL iodine flask $\xrightarrow[\text{0.4 g}]{\text{product}}$ $\xrightarrow[\text{20 mL}]{5\ \text{mol·L}^{-1}\ \text{NaOH}}$ $\xrightarrow[\text{3-4 grains}]{\text{boiling chip}}$ heat to boiling for 25 min $\longrightarrow$ add water occasionally $\longrightarrow$ cool to RT $\xrightarrow[\text{0.8 g}]{\text{KI}}$ gently swirl for 1 min $\xrightarrow[\text{20 mL}]{6\ \text{mol·L}^{-1}\ \text{HCl}}$ keep in the dark for 25 min $\xrightarrow[\text{70 mL}]{H_2O}$ titrate fast with $Na_2S_2O_3$ until light orange $\xrightarrow[\text{2 mL}]{\text{starch}}$ deep blue $\longrightarrow$ titrate slowly with $Na_2S_2O_3$ until solution turns pink $\longrightarrow$ record the volume $\longrightarrow$ repeat two more times

Note:

(1) Ensure the iodine flask is uncovered and several boiling chips are added before heating.

(2) Pour a large amount of water into the iodine flask immediately after taking it out of the dark.

(3) Titrate quickly and swirl gently before adding the starch solution.

(4) Add the starch near the endpoint to prevent the absorbance of I_2.

(5) Titrate slowly and swirl vigorously after adding the starch solution to release the absorbed I_2.

6. Determination of the Ion Configuration of Coordination Compounds

(1) Use a 100 mL volumetric flask to prepare 100 mL of 1.0×10^{-3} mol·L^{-1} product solution. **Note:** First, calculate the required amount of product, then weigh it using an analytical balance.

[221] 校准电极常数

(2) Measure the conductivity of the solution using a conductivity meter. **Note:** Calibrate the electrode coefficient[221] and temperature before measuring.

(3) Determine the ion configurations of the coordination compounds using the data in Table 5-11.

7. Measurement of the Crystal Field Splitting Energy of Coordination Compounds

Design the procedures needed to accomplish this measurement before coming to the lab, and then carry out the measurement during the lab

[222] 分光光度计

session. **Note:** A spectrophotometer[222] will be used for this measurement.

Data Treatment and Analysis

(1) Calculate the theoretical mass and percent yield of each product.

(2) Calculate the cobalt content in each product and analyze the measurement error.

(3) Calculate the accurate concentration of $Na_2S_2O_3$ solution and

analyze the measurement error.

(4) Report the ion configuration of each product and analyze the measurement error.

(5) Calculate the crystal field splitting energy of each product and explain the source of the difference obtained[223].

[223] 计算每种产物的晶体场分裂能并解释为何不同

Safety and Waste Disposal

(1) Wear a lab coat and safety goggles at all times in the laboratory. Wear gloves when handling hot glassware. If any chemicals splash into the eyes or onto the skin, wash with water immediately.

(2) NaOH, HCl and $NH_3 \cdot H_2O$ are corrosive chemicals. H_2O_2 is highly corrosive and oxidized[224]. Handle them with care.

[224] 腐蚀性和氧化性

(3) $CoCl_2 \cdot 6H_2O$ is a pink or red crystal which turns blue after losing water molecules[225]. It is categorized as a highly toxic chemical which can cause cancer and skin sensitivity.

[225] 失去水分子

(4) Do not pour the acidic, alkaline or any other waste solutions down the sink. First treat them.

Post-lab Questions

(1) If the conductivity results of the products are higher or less than the theoretical ones, give an explanation.

(2) Which λ_{max} is highest among the three cobalt (III) coordination compounds synthesized in this experiment? You should take the spectrochemical series of ligands into account[226].

[226] 需考虑配体的光谱化学序

Exploring Experiments

When cobalt-amine coordination compounds are heated with an excess strong alkaline solution, NH_3 gas will evaporate. If this gas is absorbed by a certain amount of H_2SO_4 solution, the excess H_2SO_4 can be titrated using a standard NaOH solution. This process allows us to obtain the ammonia (NH_3) content and ammonia molecular numbers in the cobalt coordination compound.

扫一扫　　视频 5-14　三种钴氨配合物的制备及表征(讲解)
　　　　　　视频 5-15　三种钴氨配合物的制备(实验)
　　　　　　视频 5-16　钴氨配合物的钴含量测定(实验)

(曾秀琼编写)

Expt. 18   Determination of Cement Composition
水泥组分测定

Objectives

(1) Understand the principles of determining cement composition.

[227] 重量分析法

(2) Learn the theory and technique of gravimetric analysis[227].

[228] 配位滴定

(3) Determine the contents of main metal oxides using complexometric titration[228].

(4) Determine the content of SiO_2 using gravimetric analysis.

(5) Develop comprehensive abilities for handling complex samples.

Principles

Key terms: complexometric titration, gravimetric analysis, constant weight, acidity control, masking method[229].

[229] 恒重、酸度控制、掩蔽法

[230] 水硬水泥

The first record of use of hydraulic cement[230] dates back to ancient Greece and Rome. The first cement patent was granted in 1824 in England, and the first modern cement production began in 1843 near London. Today, cement is indispensable for most concrete[231] structures. Every year, a significant amount of Ordinary Portland Cement (OPC) is produced and used in the construction of buildings, roads, highways, and more.

[231] 混凝土

[232] 固型和硬化

Cement, also known as "hydraulic cement", is capable of setting and hardening[232] when mixed with water. The basic ingredients of raw cement include concrete, mortar, and gypsum[233], which consists of a mixture of calcium (Ca), silicon (Si), and aluminum (Al) oxides. According to Chinese National Standards (GB)[234], cement is classified into several types: sodium silicate cement[235] (Portland cement, cement clinker), ordinary sodium silicate cement, slag cement[236], volcanic pumice cement[237], and fly-ash cement[238].

[233] 砂浆和石膏

[234] 国家标准(GB)

[235] 硅酸钠水泥

[236] 矿渣水泥

[237] 火山泥水泥

[238] 粉煤灰水泥

[239] 水泥熟料

Cement clinker[239] is produced by burning raw cement at 1400°C. The composition of cement clinker includes the following percentages: SiO_2 (18%-24%), Fe_2O_3 (2.0%-5.5%), Al_2O_3 (4.0%-9.5%), CaO (60%-67%) and MgO (< 6.0%).

In this experiment, the methods used to determine the compositions of cement adhere to the Chinese National Standard (GB/T 176—2017).

1. Treatment of Cement Sample

[240] 碱性氧化物

The content of alkaline oxides[240] in cement clinker exceeds 60%. These alkaline oxides are represented by compounds such as $3CaO \cdot SiO_2$, $2CaO \cdot SiO_2$, $3CaO \cdot Al_2O_3$, and $4CaO \cdot Al_2O_3 \cdot Fe_2O_3$. As a result, cement clinker readily decomposes into insoluble silicic acid[241] and soluble chlorides[242] when treated with hydrochloric acid (HCl).

[241] 不溶性硅酸
[242] 可溶性氯化物

2. Principle of SiO₂ Content Determination

The SiO₂ content is determined using gravity analysis. After decomposition by hydrochloric acid (HCl), the cement sample is heated at 100-110°C to convert the silicic acid solution into a hydrogel[243], making it easier to precipitate and filter. Ammonium chloride (NH₄Cl) is added to enhance the dehydration of silicic acid[244].

[243] 水凝胶

[244] 硅酸脱水

$$NH_4Cl + H_2O \rightleftharpoons NH_3 \cdot H_2O + HCl$$

Through the processes of filtration, washing, calcination at 100-150°C, and burning (igniting) at 950-1000°C to constant weight[245], the hydrogel is converted into SiO₂ solid. Finally, the SiO₂ solid is weighed to determine the SiO₂ content.

[245] 过滤、洗涤、灼烧并煅烧至恒重

$$H_2SiO_3 \cdot nH_2O \xrightarrow{110°C} H_2SiO_3 \xrightarrow{950\text{-}1000°C} SiO_2$$

For more details about gravimetric analysis, watch the related video.

3. Principle of Metal Oxides Content Determination

Fe, Al, Ca, Mg composition exist in the filtrate[246] as Fe^{3+}, Al^{3+}, Ca^{2+}, Mg^{2+} ions, which can form stable complex with EDTA. These complex have distinct stability constants[247] (Fig. 5-7), allowing them to be titrated separately if the acidity is controlled or if masking agents[248] are used.

[246] 滤液

[247] 稳定常数
[248] 掩蔽剂

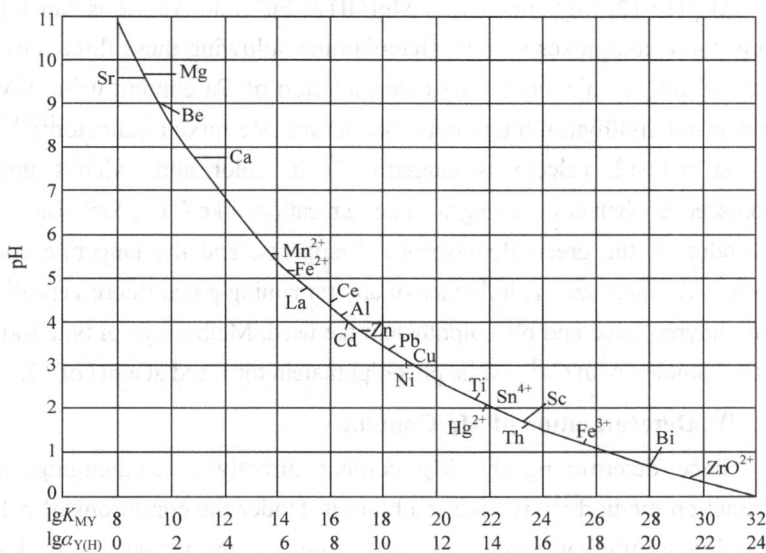

Fig. 5-7 Minimum pH needed

For more details about complexometric titration, refer to Expt. 7.

Ⅰ. Determination of Fe Content

Under the conditions of pH 1.5-2.5 and 60-70°C, the Fe content is

[249] 滴定剂

[250] 磺基水杨酸

[251] 红棕色氢氧化物

[252] 铁含量偏高而铝含量偏低

[253] 返滴定法

[254] 用三乙醇胺掩蔽

[255] 钙黄绿素-甲基百里香酚蓝-酚酞,作为混合指示剂

[256] 橘黄色

[257] 为了减小残余绿色荧光(对终点判断)的干扰

[258] 差减法

[259] 三乙醇胺-酒石酸钾钠为掩蔽剂

determined using a standard EDTA solution as the titrant[249] and 5-sulphosalicylic acid[250] as the indicator. At the endpoint, the solution's color changes from purplish-red to bright yellow. Acidity and temperature are the key factors that influence this process.

If the pH drops below 1.5, the determination result will be lower. If the pH exceeds 3.0, Fe^{3+} forms a reddish brown hydroxide[251], preventing detection of the endpoint, which also leads to a lower determination result.

If the temperature exceeds 70℃, Al^{3+} can also form a complex with EDTA, leading to an overestimation of Fe content and underestimation of Al content[252]. If the temperature is below 70℃, the complexation reaction slows down, and no accurate endpoint can be detected.

II. Determination of Al Content

Due to the slow reaction speed of Al^{3+} with EDTA, the back titration method[253] is used to determine the Al content, with a standard $CuSO_4$ solution as the titrant and PAN as the indicator. The color at the endpoint should change from yellow to bright purple.

Because the amounts of the Cu-EDTA (a deep blue complex) can effect the detection of the endpoint, the volume of excess standard EDTA solution should be controlled, using 10-15 mL in this experiment.

III. Determination of Ca Content

At pH >12, Mg^{2+} ion forms $Mg(OH)_2$, Fe^{3+} and Al^{3+} ions can form more stable complexes with triethanolamine, allowing these three ions to be masked[254]. This enable the determination of Ca content using CMP (calcein-methylthymol blue-phenolphthalein) as a mixed indicator[255].

At pH >12, calcein is tangerine[256] in color and exhibits green fluorescence when complexing with certain cations like Ca^{2+}, Sr^{2+}, Ba^{2+}. At the endpoint, the green fluorescence disappears, and the tangerine color appears. To minimize the influence of any remaining green fluorescence[257], methylthymol blue and phenolphthalein are used. Methylthymol blue forms a blue complex with Ca^{2+}, while phenolphthalein turns red at a pH of 12.

IV. Determination of Mg Content

Since determining the Mg content directly is challenging, the subtraction method[258] is used to obtain it. Under the conditions of pH = 10 and using triethanolamine and potassium sodium tartrate as masking agents[259] for Fe^{3+} and Al^{3+}, the combined content of Ca and Mg can be determined using K-B as a mixed indicator.

Pre-lab Questions

Think about the following seven questions and then answer four of them that are difficult for you.

(1) What are the functions of NH_4Cl?

(2) SiO_2 exits in the form of amorphous precipitate[260] in this experiment. Describe briefly the conditions to form an amorphous precipitate.

(3) What are the compositions of CMP? Describe the function of each composition.

(4) What are the differences between quantitative filter paper and qualitative filter paper[261]?

(5) What is the definition of constant weight? How is constant weight performed?

(6) Describe the principles behind and basic operations of gravity filtration[262].

(7) Describe the requirements for an effective complexometric titration and a separate complexometric titration.

Apparatus and Reagents

Apparatus: analytical balance, hot plate, Muffle furnace, desiccator[263], burette (25 mL), volumetric flask (250 mL), Erlenmeyer flask (3×150 mL), pipette (20 mL, 10 mL), long neck glass conical funnel, spot plate (black), watch glass, ceramic crucible with lid, crucible tongs (long, short)[264], quantitative filter paper.

Reagents: cement (s), NH_4Cl (s), HCl (concentrated, 6 $mol·L^{-1}$), HNO_3 (concentrated), $NH_3·H_2O$ (1+1), NaOH (6 $mol·L^{-1}$), KOH (20%), standard EDTA solution (0.05 $mol·L^{-1}$), standard $CuSO_4$ solution (0.01 $mol·L^{-1}$), $AgNO_3$ (0.1 $mol·L^{-1}$), triethanolamine (1+2), potassium sodium tartrate (10%), HAc-NaAc buffer (pH 4.3), NH_3-NH_4Cl buffer (pH 10), bromcresol green[265] (0.05%, indicator), 5-sulphosalicylic acid (10%, indicator), PAN (0.3%, indicator), K-B (s, indicator), CMP (s, indicator), calconcarboxylic acid (Ca-indicator)[266].

Procedures

1. Dissolution of Cement Sample

[260] 无定形沉淀

[261] 定量滤纸和定性滤纸

[262] 常压过滤

[263] 马弗炉、干燥器

[264] 长颈漏斗、点滴板、表面皿、带盖坩埚、坩埚钳

[265] 溴甲酚绿

[266] 钙羧酸(钙指示剂)

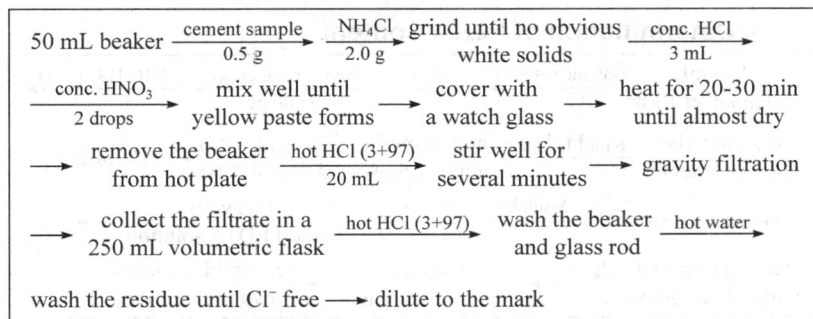

Note:

(1) Grind the cement and NH_4Cl solids with the flattened end of a glass rod until no obvious white solids remain. Add concentrated HCl and HNO_3, mix well, and heat the mixture until almost dry[267].

(2) Prepare diluted HCl (3+97) by adding 3 mL (or 3 droppers) of concentrated HCl to 100 mL of boiling water. Prepare this solution just before use.

(3) For better and quicker gravity filtration, prepare a water column in the glass funnel[268].

(4) Wash the mixture 2-3 times with hot HCl (3+97) followed by over 10 washes with hot water, each time adding 10 mL and transferring the supernatant to filter paper (leaving the solids in the beaker)[269].

(5) When the volumetric flask is about 2/3 full of the filtrate, remove the funnel and add one drop of the filtrate to a black spot plate. Then, add one drop of $AgNO_3$ to check if any Cl^- ion remains[270].

2. Determination of SiO₂ Content

filter paper and residue	→	a weighed porcelain crucible	→	calcinate on a hot plate until filter paper turns to ashes
→ ignite at 850℃ for 30 min	→	keep in a desiccator until cooling to RT	→	weigh and record the mass
→ repeat igniting and weighing until constant weight				

Note:

(1) First, calcinate the filter paper on a hot plate until it turns to ash[271].

(2) Use crucible tongs to handle the crucible carefully, ensuring it is not tilted[272]. If the crucible flips, the procedure must be repeated, which is tedious and time-consuming.

(3) When cooling to room temperature (RT), leave one corner of the desiccator uncovered at the start[273].

(4) Ensure that the cooling to RT is the same for each sample. Weigh the sample using the same analytical balance to minimize the system errors[274].

3. Determination of Fe₂O₃ Content

150 mL Erlenmeyer flask	sample solution 20.00 mL →	H₂O 40 mL →	bromocresol green 2-4 drops →	NH₃·H₂O (1+1)
add dropwise until pale green	6 mol·L⁻¹ HCl →	add dropwise until yellow	6 mol·L⁻¹ HCl 3 drops in excess →	pH 1.8-2.0
→ heat to 60-70℃	5-sulphosalicylic acid 5-8 drops →	titrate with standard EDTA solution →		
(color) from reddish purple to bright yellow	→	read and record the finial volume	→	repeat this process two more times

[267] 用平头玻璃棒搅拌至无明显白色颗粒。加入浓酸搅匀后慢热至近干

[268] 常压过滤，玻璃漏斗做好水柱

[269] 先用热的稀盐酸洗涤，再用热水洗涤，每次只转移上清液到滤纸上

[270] 取下漏斗，滴一滴滤液到黑色点滴板上，检查是否还有 Cl^-

[271] 滤纸先在电炉上灰化

[272] 用坩埚钳小心操作坩埚，切勿倒翻

[273] 起初干燥器不能完全密封

[274] 减小系统误差

Note:

(1) Adjust the acidity carefully. Bromocresol green solution is yellow at pH < 3.8 and green at pH > 5.4.

(2) Heat the solution to 60-70ºC, but not to boiling[275]. Do not use a thermometer to check the temperature. When significant water vapor forms around the neck of the Erlenmeyer flask, the temperature should be around 60-70ºC.

(3) Ensure the temperature at the endpoint is above 60ºC, as it can be difficult to detect the endpoint otherwise.

(4) Titrate slowly in this step, as the required volume of standard 0.01 mol·L^{-1} EDTA solution[276] is small (about 2 mL).

(5) Keep the titrated solution for the next step[277].

4. Determination of Al₂O₃ Content

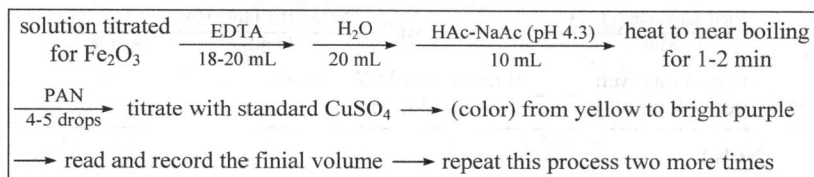

solution titrated for Fe₂O₃ $\xrightarrow[\text{18-20 mL}]{\text{EDTA}}$ $\xrightarrow[\text{20 mL}]{\text{H}_2\text{O}}$ $\xrightarrow[\text{10 mL}]{\text{HAc-NaAc (pH 4.3)}}$ heat to near boiling for 1-2 min

$\xrightarrow[\text{4-5 drops}]{\text{PAN}}$ titrate with standard CuSO₄ ⟶ (color) from yellow to bright purple

⟶ read and record the finial volume ⟶ repeat this process two more times

Note:

(1) Use a burette to accurately add[278] 18-20 mL of standard 0.01 mol·L^{-1} EDTA solution.

(2) Maintain a high temperature at all times[279], as a lower temperature makes it difficult to detect the endpoint.

5. Determination of CaO Content

Method 1: Follow the procedures in GB using CMP as the indicator.

It is a little difficult to detect the endpoint using this method.

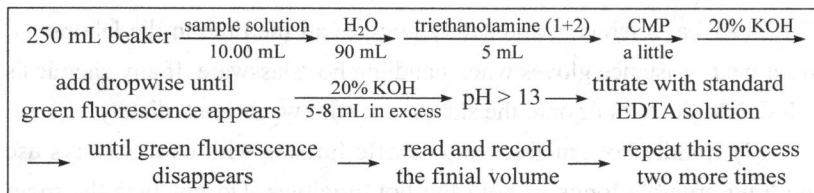

250 mL beaker $\xrightarrow[\text{10.00 mL}]{\text{sample solution}}$ $\xrightarrow[\text{90 mL}]{\text{H}_2\text{O}}$ $\xrightarrow[\text{5 mL}]{\text{triethanolamine (1+2)}}$ $\xrightarrow[\text{a little}]{\text{CMP}}$ 20% KOH

add dropwise until green fluorescence appears $\xrightarrow[\text{5-8 mL in excess}]{\text{20\% KOH}}$ pH > 13 ⟶ titrate with standard EDTA solution

⟶ until green fluorescence disappears ⟶ read and record the finial volume ⟶ repeat this process two more times

Note:

(1) Use a beaker instead of an Erlenmeyer flask.

(2) Observe the color change and disappearance of green fluorescence from the top of beaker[280].

[275] 不能加热至沸

[276] 若EDTA标准溶液浓度为 0.05 mol·L^{-1}，需先准确稀释至 0.01 mol·L^{-1} (移取 50.00 mL 至 250 mL 容量瓶，用水定容)

[277] 将滴定后的溶液用于下一步

[278] 用滴定管准确加入

[279] 全程保持高温

[280] 从烧杯上部观察颜色变化和绿色荧光消失

Method 2: Use calconcarboxylic acid (Ca-indicator) as the indicator.

This method makes it easier to detect the endpoint.

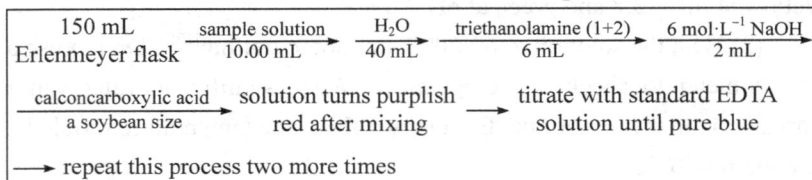

150 mL Erlenmeyer flask	→ sample solution 10.00 mL	→ H_2O 40 mL	→ triethanolamine (1+2) 6 mL	→ 6 mol·L^{-1} NaOH 2 mL →

calconcarboxylic acid a soybean size	→ solution turns purplish red after mixing	→ titrate with standard EDTA solution until pure blue

→ repeat this process two more times

Note:

(1) Control the amount of the indicator.

(2) The solution turns pure blue at the endpoint[281].

[281] 滴定终点时，溶液
变为纯蓝色

6. Determination of MgO and CaO Contents

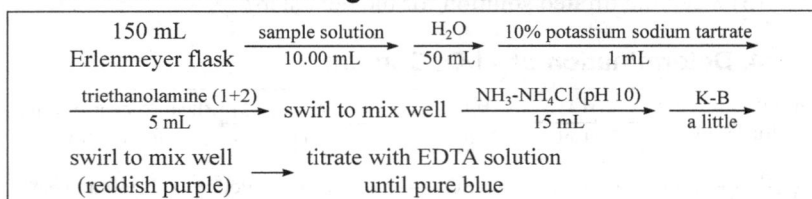

150 mL Erlenmeyer flask	→ sample solution 10.00 mL	→ H_2O 50 mL	→ 10% potassium sodium tartrate 1 mL →

→ triethanolamine (1+2) 5 mL	→ swirl to mix well	→ NH_3-NH_4Cl (pH 10) 15 mL	→ K-B a little →

swirl to mix well (reddish purple)	→ titrate with EDTA solution until pure blue

Note:

(1) Control the amount of K-B.

(2) The solution turns pure blue at the endpoint.

Data Treatment and Analysis

(1) Calculate the percentage of each oxide determined, and present the data in separate tables[282]. Then, calculate the total content of these oxides.

(2) Analyze whether this cement sample meets the requirements of the relevant cement standards[283].

[282] 每种组成测定结果
单独列一个表格
[283] 分析是否满足相关
水泥标准

Safety and Waste Disposal

(1) Wear a lab coat and safety goggles all the time in the laboratory. Wear heat resistance gloves when handling hot glassware. If any chemicals splash into the eyes or onto the skin, wash with water immediately.

(2) Handle hot crucibles and Muffle furnace with care. Always use the long crucible tongs to hold the hot crucibles. Do not face the inner surface of hot Muffle furnace towards any people[284].

[284] 不能让高温马弗炉
炉门内侧朝向任何人

(3) Concentrated HCl, HNO_3 and NH_3·H_2O are strongly corrosive and volatile, perfume them in the fume hood. Be careful when handling all of these solutions.

(4) Never pour the waste solutions down the sink, treat them first.

Post-lab Questions

(1) What are the key points to obtain an accurate result of SiO_2 content?

(2) When determining the CaO content, why is triethanolamine added ahead of KOH?

Exploring Experiments

Design proposals to complete the following tasks.

(1) Determine the content of soluble SiO_2.

(2) Determine the content of the volatile compositions.

(3) Determine the content of metal oxides using other methods.

| 扫一扫 | 视频 5-17   重量分析法概述 |
| | 视频 5-18   水泥组分测定(讲解) |

| 扫一扫 | 视频 5-19   水泥样品处理和 SiO_2 含量测定(实验) |
| | 视频 5-20   水泥样品中铁、铝、钙、镁含量测定(实验) |

(曾秀琼编写)

Expt. 19   Preparation and Determination of Three Kinds of Sodium Phosphate Hydrates
三种磷酸钠盐水合物的制备及测定

Objectives

(1) Master the principles of polyprotic acid and base titrations[285].   [285] 多元酸碱滴定

(2) Learn the properties and preparation methods of three sodium phosphate hydrates[286].   [286] 磷酸钠盐水合物

(3) Design the procedures to determine the purity of the products.

Principles

Key Terms: alcohol-precipitation[287], weak electrolyte, hydrolysis[288], polyprotic acid (base) titration, potentiometric titration[289].

[287] 醇析
[288] 水解
[289] 电位滴定

Trisodium phosphate (Na_3PO_4)[290], disodium hydrogen phosphate (Na_2HPO_4)[291], and sodium dihydrogen phosphate (NaH_2PO_4)[292] are important industrial materials used as water softener, cleaning agents, metal antirust, and buffer solutions. These salts can form various white or colorless hydrated crystals, which are efflorescent[293] in air, highly soluble in water, but insoluble in organic solvents.

[290] 磷酸钠
[291] 磷酸氢二钠
[292] 磷酸二氢钠
[293] 易风化

[294] 十水合物、六水合
物、一水合物和无水合物
[295] 七水合物

[296] 二水合物

[297] 焦磷酸盐
[298] 偏磷酸盐

[299] 磷酸根的来源

[300] 由于在水中溶解度
大，用乙醇使产品沉淀

[301] 由于滴定曲线上只
有两个化学计量点，磷酸
可作为一元酸或二元酸
滴定

[302] 课前设计实验步骤，
课中实施

(1) $Na_3PO_4 \cdot 12H_2O$, becomes a decahydrate at 55-65°C, a hexahydrate at 60-100°C, a monohydrate above 100°C, and an anhydrate[294] above 212°C. Its aqueous solution is alkaline, with a pH around 12.

(2) $Na_2HPO_4 \cdot 12H_2O$, becomes a heptahydrate[295] at 35.1°C and an anhydrate above 100°C. When carefully dried at 34°C, it converts to a dihydrate[296] (white powder). Its aqueous solution is weakly alkaline, with a pH of 8.0-11.0.

(3) $NaH_2PO_4 \cdot 2H_2O$, becomes an anhydrate above 100°C, pyrophosphate[297] at 190-210°C, and metaphosphate[298] at 280-300°C. Its aqueous solution is weakly acidic, with a pH around 4.5.

In this experiment, phosphoric acid (H_3PO_4) or $Ca(H_2PO_4)_2 \cdot H_2O$ is used as the source of phosphate ions[299] (PO_4^{3-}) to prepare three types of sodium phosphate hydrate. Sodium hydroxide (NaOH) or sodium carbonate (Na_2CO_3) is used to adjust the pH of the reaction system. Due to their high solubility in water, the products are precipitated using ethanol[300].

Titration of polyprotic acids (bases) requires more attention than that of monoprotic acids (bases). Phosphoric acid (H_3PO_4) is a relatively weak acid with $pK_{a1} = 2.15$, $pK_{a2} = 7.20$ and $pK_{a3} = 12.35$. As shown in Fig. 5-8, the titration curve contains only two stoichiometric points (pH = 4.70, pH = 9.65), meaning that H_3PO_4 can be titrated either as a monoprotic or diprotic acid[301].

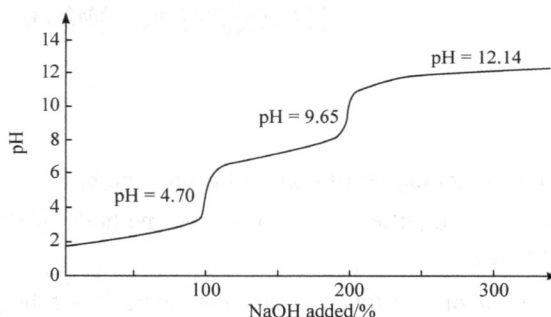

Fig. 5-8　Titration curve of H_3PO_4 with NaOH

The purity of the products can be determined using either traditional polyprotic acids (bases) titration or potentiometric titration. Design the necessary procedures before coming to the laboratory and then perform them during the laboratory session[302].

Pre-lab Questions

(1) Briefly explain why it is important to control the acidity accurately when preparing these three products.

(2) Briefly describe the principles of polyprotic acid and base titrations.

(3) Briefly summarize the basic operations of an automatic potentiometric

titrator.

Apparatus and Reagents

Apparatus: analytical balance, hot plate, magnetic stirrer[303], automatic potentiometric titrator[304], pH meter, vacuum pump, suction flask, Buchner funnel, burette (25 mL), evaporating dish[305], stainless-steel spatula[306].

Reagents: Na_2CO_3 (s), $Ca(H_2PO_4)_2 \cdot H_2O$ (s), H_3PO_4 (6 mol·L^{-1}), HCl (0.1 mol·L^{-1}), NaOH (6 mol·L^{-1}, 2 mol·L^{-1}, 0.1 mol·L^{-1}), indicators (phenolphthalein, methyl orange, methyl red, thymolphthalein[307]), pH test paper [universal(1-14), 3.8-5.4, 8.2-10.0] , anhydrous ethanol, ice.

Procedures

General notes:

(1) Use a magnetic stirrer to ensure the reaction reaches full completion. A 150 mL Erlenmeyer flask is preferable to a beaker to prevent splashing during stirring[308].

(2) If a large amount of solution needs to be evaporated, place the evaporating dish directly on a hot plate, or divide the solution between two evaporating dishes[309].

(3) When measuring pH, first use a universal pH test paper (1-14), followed by a narrow-range test paper[310] (8.2-10.0 or 3.8-5.4) for greater accuracy.

(4) Stir the product solution with a stainless-steel spatula[311] before complete precipitation.

(5) Control the amount of water added in each step to avoid prolonging evaporation time.

(6) When drying the product, use an oven or spread it on a watch glass to air-dry[312].

1. Preparation of Na$_3$PO$_4$·12H$_2$O

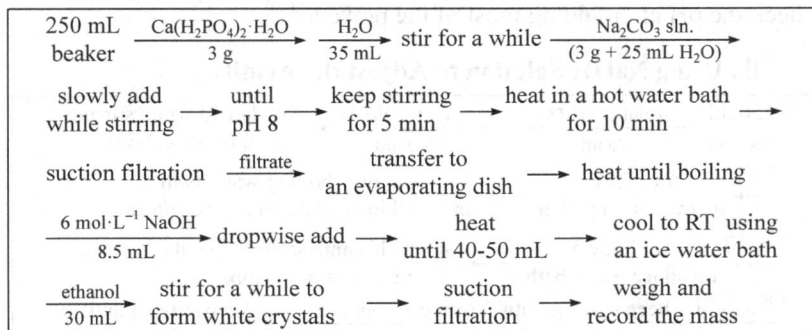

Note:

(1) Prepare the Na_2CO_3 solution by adding 3 g of Na_2CO_3 to 25 mL

[303] 磁力搅拌器
[304] 自动电位滴定仪
[305] 蒸发皿
[306] 不锈钢勺

[307] 百里酚酞

[308] 在锥形瓶中进行以避免溶液溅出

[309] 或者将溶液分在两个蒸发皿中

[310] 先用广范 pH 试纸，再用精密试纸

[311] 用不锈钢勺搅拌

[312] 置于表面皿中自然干燥

of H_2O in a 100 mL beaker. Heat until all solids dissolve completely.

(2) When aging $CaCO_3$, avoid stirring the solution[313].

[313] (热水浴)陈化时，不要搅拌溶液

2. Preparation of $Na_2HPO_4 \cdot 12H_2O$

Note:

(1) Intermittently place the evaporating dish in an ice water bath. Keep stirring and be careful not to let the solution cool too quickly to prevent agglomeration[314].

(2) If no crystals form after the solution has completely cooled, use a stainless-steel spatula to quickly rub the inner surface of the evaporating dish to induce crystallization[315].

[314] 间歇式地将蒸发皿置于冰浴上冷却并不停搅拌。不能冷却过快以防板结
[315] 快速摩擦蒸发皿内壁促使结晶

3. Preparation of $NaH_2PO_4 \cdot 2H_2O$

I. Using Na_2CO_3 Powders to Adjust the Acidity

Note: The required amount of Na_2CO_3 powder is approximately 8.5 g. Check the pH after adding most of the powder[316].

[316] 大部分粉末加完后，检测 pH

II. Using NaOH Solution to Adjust the Acidity

Note: The required amount of NaOH solution is approximately 30 mL. Check the pH after adding most of the solution.

Assignments of this Experiment

Complete all the following assignments in groups. Design the procedures (3) and (4) before coming to the lab.

(1) Prepare these three types of products mentioned above.

(2) Prepare solutions of each product and measure their pH using pH test paper.

(3) Determine the purity of each product using the traditional acid-base titration method with a burette.

(4) Determine the purity of each product using an automatic potentiometric titrator[317] and a pH meter.   [317] 自动电位滴定仪

Safety and Waste Disposal

(1) Wear a lab coat and safety goggles at all times in the laboratory. Wear gloves when handling hot glassware. If any chemicals splash into the eyes or onto the skin, wash with water immediately.

(2) NaOH, H_3PO_4, Na_3PO_4, Na_2HPO_4, NaH_2PO_4, Na_2CO_3, etc, are corrosive with varying strengths, they should be handled with care.

(3) Never pour waste acidic or alkaline solutions down the sink, neutralize them first.

Post-lab Questions

If there is a bag full with a white mixture, which is (are) one (several) of the following chemicals, how can the content and identity of the chemical be determined? $NaH_2PO_4 \cdot 2H_2O$, $Na_2HPO_4 \cdot 12H_2O$ and $Na_3PO_4 \cdot 12H_2O$.

Exploring Experiment

Potentiometric determination of the purity of products using a pH meter.

Procedures:

(1) Accurately weigh the product (or accurately pipette the sample solution, prepared by a volumetric flask) into a beaker. **Note:** First, calculate the required amount of product, then weigh it using an analytical balance.

(2) Add water to dissolve the sample completely.

(3) Place the beaker on a magnetic stirrer, insert the stirrer bar and electrode (Fig. 5-9), determine the initial pH[318].

[318] 放上搅拌子和电极，测定起始 pH

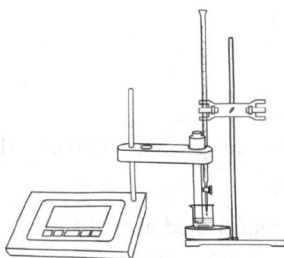

Fig. 5-9　The apparatus

(4) Calculate the required volume of $0.1\ mol\cdot L^{-1}$ HCl to reach the stoichiometric point.

(5) Add HCl to the sample solution in 5-6 mL increments (for 20 mL titrant required) or 2-3 mL increments (for 10 mL titrant required) until 1 mL before the expected stoichiometric point. Record the pH after each addition, then reduce the HCl increments to 0.2-0.3 mL.

(6) Continue the titration, recording the pH until 1 mL past the expected stoichiometric point.

Tasks:

(1) Draw the potentiometric titration curve, find the stoichiometric point.

(2) Calculate the purity of the product.

扫一扫　视频 5-21　三种磷酸钠盐水合物的制备及表征(讲解)
　　　　视频 5-22　三种磷酸钠盐水合物的制备(实验)
　　　　视频 5-23　自动电位滴定仪的使用

(蔡吉清编写)

Expt. 20　Preparation of Heteropolyacid Salts and Determination of Conversion Kinetics
杂多酸盐的制备及其转化动力学测定

Objectives

(1) Understand the principles and methods for preparing two heteropolyacid salts[319], $K_5CoW_{12}O_{40}$ and $K_6CoW_{12}O_{40}$.

[319] 杂多酸盐

(2) Understand the external redox principle[320] of metal ions in tungsten-cobalt heteropolyacid salts.

[320] 外界氧化还原机理

(3) Measure the redox reaction order[321] and rate constant[322] of $K_6CoW_{12}O_{40}$ using spectrophotometry.

[321] 反应级数
[322] 速率常数

Principles

Key terms: heteropolyacid and its structure, spectrophotometry[323], reaction kinetics[324].

[323] 分光光度法

[324] 反应动力学

Heteropolyacid are formed through the dehydration of two different oxygen-contained acids[325]. The salts resulting from the substitution of H$^+$ by metal ions[326] are called heteropolyacid salts. Both heteropolyacids and their salts are significant in inorganic chemistry. Like polyprotic acids[327], heteropolyacids exhibit similar characteristics, such as neutralization reactions[328], substitution of H$^+$ by metal ions to form salts, and substitution reactions[329] with coordinate groups. They have attracted significant attention due to their special catalytic and antiviral properties.

[325] 两种不同含氧酸的脱水

[326] H$^+$被金属离子取代

[327] 多元酸

[328] 中和反应

[329] 取代反应

The tetrahedral structures of dodectungsten (or molybdenum) heteropolyacid salts[330] have been the most extensively studied. The metal ion serves as the central atom of the complex, with polyacid anions ($W_3O_{10}^-$) acting as ligands positioned at the apexes[331] of the tetrahedron. This structure is known as the Keggin structure.

[330] 十二钨(或钼)杂多酸盐的四面体结构

[331] 配体位于顶角

In this experiment, tungsten-cobalt heteropolyacid salts[332] are prepared using Baker's method.

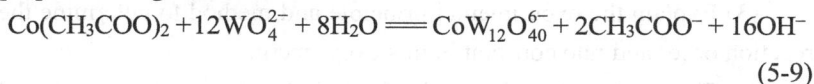

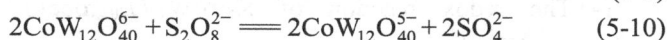

[332] 钨-钴杂多酸盐

$$Co(CH_3COO)_2 + 12WO_4^{2-} + 8H_2O = CoW_{12}O_{40}^{6-} + 2CH_3COO^- + 16OH^-$$
$$(5\text{-}9)$$

$$2CoW_{12}O_{40}^{6-} + S_2O_8^{2-} = 2CoW_{12}O_{40}^{5-} + 2SO_4^{2-} \qquad (5\text{-}10)$$

The redox reactions of heteropolyacids and their salts follow either the external reaction mechanism or the internal reaction mechanism[333]. In the external mechanism, electrons are transferred between different molecules, while in the internal mechanism, electrons are transferred through atoms or groups within a single molecule.

[333] 内界反应机理

The $CoW_{12}O_{40}^{5-}$ redox reaction follows the external reaction mechanism. The Co(III) contained heteropolyacids are strong oxidizers[334] and can be reduced into $CoW_{12}O_{40}^{6-}$. In this process, both the oxidation state of cobalt and the charge of the heteropolyacid ion are altered.

[334] 含 Co(III)的杂多酸是强氧化剂

During the $K_5CoW_{12}O_{40}$ reduction by KSCN, the chemical equation can be written as

$$2K_5CoW_{12}O_{40} + 2KSCN = 2K_6CoW_{12}O_{40} + (SCN)_2 \qquad (5\text{-}11)$$

When the amount of reductant KSCN is excessive, the reaction rate can be written as Equation (5-12)

$$-\frac{dc}{dt} = k_{obs}c^n \qquad (5\text{-}12)$$

where c is the concentration of $K_5CoW_{12}O_{40}$.

If the plot of $-\ln c$ versus t is a straight line, it indicates that $n = 1$, meaning the redox reaction is a first-order reaction. The slope is the apparent rate constant k_{obs}[335].

There is a distinct absorption peak around 388 nm in the UV-Vis spectrum[336] of $K_5CoW_{12}O_{40}$. According to the Lambert-Beer law, the concentration of $K_5CoW_{12}O_{40}$ can be determined by measuring the absorbance at 388 nm.

During the redox reaction, $K_5CoW_{12}O_{40}$ is not completely reduced, some remains in the system. The absorbance of this remaining $K_5CoW_{12}O_{40}$ is denoted as A_∞, while the absorbance of reduced $K_5CoW_{12}O_{40}$ is denoted as $(A-A_\infty)$. When changing the reductant concentration to obtain a series of k_{obs}, the reaction order of the reductant can be determined by plotting the k_{obs} versus the reductant concentration as a straight line. The rate constant k can be obtained from the slope of the line[337].

[337] (本段要点)$K_5CoW_{12}O_{40}$ 不能被全部还原，残留在溶液的吸光度标为 A_∞，因此被还原的吸光度标为 $(A-A_\infty)$。作图得到直线的斜率，即为速率常数 k

[338] 基元反应和反应级数

Pre-lab Questions

(1) Explain the definition and structure of heteropolyacids. Draw the Keggin structure of $K_5CoW_{12}O_{40}$.

(2) Explain the definition of elementary reaction and reaction order[338].

(3) Explain the experimental principle and method to determine the reaction order and rate constant in this experiment.

(4) The redox reaction of $K_5CoW_{12}O_{40}$ obeys the external mechanism. Why?

Apparatus and Reagents

Apparatus: UV-Vis spectrophotometer, vacuum pump, Buchner funnel, suction flask, beaker, graduated cylinder, volumetric flask (50 mL, 100 mL), colorimetric tube (25 mL), pipette (5 mL, 10 mL), pH test paper (1-14, 5.5-9.0).

Reagents: $Na_2WO_4 \cdot 2H_2O$(s), $Co(CH_3COO)_2 \cdot 4H_2O$(s), $K_2S_2O_8$(s), KCl(s), KSCN(s), H_2SO_4 (3 mol·L^{-1}), HAc (glacial acetic acid), ethanol.

Procedures

1. Preparation of $K_6CoW_{12}O_{40} \cdot 16H_2O$

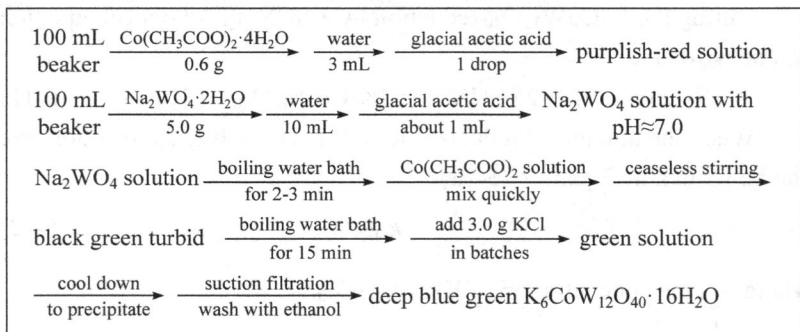

Note:

(1) When adjusting the pH, reduce the amount of HAc (glacial acetic acid) at higher room temperatures[339]. Some white floccule[340] may appear and then disappear during this process.

(2) Blue-purple powder may form at the bottom of the beaker, stir with a glass rod to aid dissolution. Add some water to prevent evaporation and drying[341].

(3) When adding KCl solids, stir constantly to prevent the formation of oil-like substances[342].

(4) Two distinct layers should form: the upper layer will be blue, and the lower layer will consist of blue and green powders.

[339] 温度高时，(因溶解度变大)减少冰醋酸的用量

[340] 白色絮状物

[341] 适量补水以防蒸干

[342] 加入 KCl 时，不停搅拌以防产生油状物

2. Preparation of $K_5CoW_{12}O_{40} \cdot 20H_2O$

I. Formation of $K_6CoW_{12}O_{40} \cdot 16H_2O$

100 mL beaker	$\xrightarrow[\text{1.2 g}]{Co(CH_3COO)_2 \cdot 4H_2O}$	$\xrightarrow[\text{6-7 mL}]{water}$	$\xrightarrow[\text{1 drop}]{glacial\ acetic\ acid}$ purplish-red solution
100 mL beaker	$\xrightarrow[\text{10.0 g}]{Na_2WO_4 \cdot 2H_2O}$	$\xrightarrow[\text{20 mL}]{water}$	$\xrightarrow[\text{about 2 mL}]{glacial\ acetic\ acid}$ Na_2WO_4 solution with pH≈6.5

Na_2WO_4 solution $\xrightarrow[\text{for 2-3 min}]{boiling\ water\ bath}$ $\xrightarrow[\text{mix quickly}]{Co(CH_3COO)_2\ solution}$ $\xrightarrow{ceaseless\ stirring}$

black green turbid $\xrightarrow[\text{for 20 min}]{boiling\ water\ bath}$ $\xrightarrow[\text{in batches}]{add\ 6.6\ g\ KCl}$ green solution

$\xrightarrow[\text{to precipitate}]{cool\ down}$ $\xrightarrow[\text{wash with filtrate}]{suction\ filtration}$ deep blue green $K_6CoW_{12}O_{40} \cdot 16H_2O$

Note: The same instruction from "1. Preparation of $K_6CoW_{12}O_{40} \cdot 16H_2O$" can be applied here.

II. Formation of $K_5CoW_{12}O_{40} \cdot 20H_2O$

100 mL beaker	$\xrightarrow{K_6CoW_{12}O_{40} \cdot 16H_2O}$	$\xrightarrow[\text{8 mL}]{3\ mol \cdot L^{-1}\ H_2SO_4}$	$\xrightarrow[\text{filter off the solids}]{stir\ to\ dissolve}$

$K_6CoW_{12}O_{40}$ solution $\xrightarrow[\text{for 3-5 min}]{boiling\ water\ bath}$ $\xrightarrow[\text{in batches}]{add\ 2.0\ g\ K_2S_2O_8}$ golden yellow solution

$\xrightarrow[\text{until more no bubbles}]{boiling\ water\ bath}$ keep the volume at 15-20 mL $\xrightarrow[\text{to precipitate}]{cool\ down}$ $\xrightarrow[\text{wash with filtrate and water}]{suction\ filtration}$

golden $K_5CoW_{12}O_{40} \cdot 20H_2O$

Note:

(1) There will be many insoluble solids after treatment with sulfuric acid.

(2) When adding $K_2S_2O_8$ solids, amounts of gas bubbles will come out, which make the solution cling to the wall of the container. Add water to prevent the solution from drying due to evaporation and to flush the clinging solution back into the reaction system[343].

[343] 大量刺激性气泡的产生导致溶液悬挂在内壁上，补水时将其冲下

(3) After adding $K_2S_2O_8$, place the beaker in a boiling water bath and stirred for a few minutes until a golden-yellow solution forms. If the color doesn't change after adding 2.0 g of $K_2S_2O_8$, continue adding more $K_2S_2O_8$ in small batches, with a total dosage not exceeding 3.0 g[344].

(4) The liquids clinging to the inner wall of the beaker should be golden yellow. Add water to prevent drying from evaporation. If the liquids are noticeably green, oxidation is incomplete, continue heating and stirring.

III. Recrystallization[345] of $K_5CoW_{12}O_{40} \cdot 20H_2O$

Dissolve 5 g of crude[346] $K_5CoW_{12}O_{40} \cdot 20H_2O$ in 5-6 mL of hot water, maintain the temperature at 50-60°C (a higher temperature may cause complete dissolution). Evaporate some water, then slowly cool down[347] the solution to form golden-yellow needle-like $K_5CoW_{12}O_{40} \cdot 20H_2O$ crystals[348].

IV. Formation of Large $K_5CoW_{12}O_{40} \cdot 20H_2O$ Crystals

Dissolve 5 g of crude $K_5CoW_{12}O_{40} \cdot 20H_2O$ in 8 mL of hot water to form a clear solution[349]. Allow the solution to cool and let it sit undisturbed for 3-4 days. Large columnar crystals[350] will then formed.

3. Kinetic Measurement of $K_5CoW_{12}O_{40} \cdot 20H_2O$ Reduction

I. Preparation of $K_5CoW_{12}O_{40}$ Solution

100 mL beaker → $K_5CoW_{12}O_{40} \cdot 20H_2O$ → add little water to dissolve → transfer to a 100 mL volumetric flask		

4.0×10^{-4} mol·L^{-1} $K_5CoW_{12}O_{40}$ solution

100 mL beaker → NaSCN → add little water to dissolve → transfer to a 50 mL volumetric flask dilute to the mark → 1.00×10^{-1} mol·L^{-1} SCN$^-$ solution

transfer 1.00, 1.50, 2.00, 3.00 mL into four 25-mL colorimetric tubes → 4.00×10^{-3}, 6.00×10^{-3}, 8.00×10^{-3} and 12.0×10^{-3} mol·L^{-1} SCN$^-$ solution

Note: First, calculate the required mass of $K_5CoW_{12}O_{40} \cdot 20H_2O$ ($M_r = 3460.0$) and NaSCN ($M_r = 81.07$) before proceeding.

II. Measuring the Absorbance of Solutions Changing with Time

four clean and dry 25 mL colorimetric tubes → 10.00 mL $K_5CoW_{12}O_{40}$ solution 10.00 mL SCN$^-$ solution → measure the absorbance at 388 nm

record the data in Table 5-12

400 mL beaker → the above four colorimetric tubes with the remaining solution → boiling water bath for 24 h

record the absorbance as A_∞

Note:

(1) Mix 10.00 mL of $K_5CoW_{12}O_{40}$ and 10.00 mL of KSCN solution of different concentrations in four 25 mL colorimetric tubes. Simply mix them together without diluting to the mark[351].

(2) The absorbance of the mixed solution should initially be above 0.2.

[344] 加完 $K_2S_2O_8$ 后，需沸水浴加热并搅拌直至溶液变成金黄色。若颜色不变，可适量补加 $K_2S_2O_8$，但总量不超过 3.0 g
[345] 重结晶
[346] 粗产物
[347] 慢慢冷却
[348] 针状晶体
[349] 澄清溶液
[350] 柱状大晶体
[351] 只需混合均匀，不用定容

(3) If using multiple cuvettes, check their initial absorbance by filling each with water, using the first one as the reference solution[352].

(4) The absorbance should decrease over time.

(5) For A_∞ measurement, gently seal the colorimetric tube to prevent water evaporation, which could alter the solution concentration[353].

Data Treatment and Analysis

(1) Calculate the yield of $K_6CoW_{12}O_{40}$ and $K_5CoW_{12}O_{40}$.

(2) Fill the A values in Table 5-12.

Table 5-12   Absorbance at different times

[SCN⁻] /(mol·L⁻¹)	t/min	0	8	16	24	32	40	48	∞	k_{obs} /min⁻¹
4.0×10⁻³	A									
	$\ln(A-A_\infty)$									
6.0×10⁻³	A									
	$\ln(A-A_\infty)$									
8.0×10⁻³	A									
	$\ln(A-A_\infty)$									
12.0×10⁻³	A									
	$\ln(A-A_\infty)$									

(3) When the concentration of SCN⁻ is 4.00×10^{-3}, 6.00×10^{-3}, 8.00×10^{-3}, 12.0×10^{-3} mol·L⁻¹, plot $\ln(A-A_\infty)$ to t to obtain a series of k_{obs} values.

(4) Plot k_{obs} to the concentration of SCN⁻ to obtain the reaction order and rate constant k.

Safety and Waste Disposal

(1) Wear a lab coat and safety goggles at all times in the laboratory. Wear gloves when handling hot glassware. If any chemicals splash into the eyes or onto the skin, wash with water immediately.

(2) Na_2WO_4 is not harmful, but it can cause skin irritation, eye irritation, and respiratory irritation[354].

(3) $K_2S_2O_8$ is a colorless or white crystal with strong oxidation which can cause combustion when rubbed or mixed with organic matter[355]. It can irritate the eyes, nose, throat and so on. In case of contact with eyes or skin, immediately wash with water. In case of accidental inhalation, immediately leave, go to fresh air and keep the airway unobstructed[356]. In case of accidental ingestion, quickly drink enough warm water to vomit.

(4) Never pour the oxidant or reductant waste solution down the sink, react with each other first. Never pour the acidic or alkaline waste solution down the sink, neutralize first.

[352] 需测定几个比色皿的原始吸光度以检查配套

[353] 比色管需轻轻盖住，否则水分蒸发会改变浓度

[354] 刺激呼吸道

[355] 当摩擦或者与有机物混合时，会引起燃烧

[356] 保持气道通畅

Post-lab Questions

(1) The absorbance of the mixed solution should be above 0.2 at the beginning. Analyze the reason if you get a low absorbance.

(2) The absorbance of the mixed solution should decrease as time goes on. Analyze the reason if the absorbance does not change or rises.

Exploring Experiments

(1) How can large high quality $K_5CoW_{12}O_{40} \cdot 20H_2O$ crystals be obtained through improving the experimental conditions or methods?

(2) How can activation energy of the $K_5CoW_{12}O_{40} \cdot 20H_2O$ redox reaction be determined? Try to complete an experimental design.

扫一扫　视频 5-24　杂多酸盐的制备及其速率常数测定(讲解)
　　　　　视频 5-25　杂多酸盐的制备及其速率常数测定(实验)

(何桂金编写)

APPENDIX

Appx. 1 Illustration of Common Glassware and Apparatus
常用玻璃器皿和仪器插图

Erlenmeyer flask	Beaker	Narrow-mouth bottle	Wide-mouth bottle	Dropper bottle	Mortar & pestle	Evaporating dish

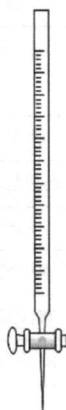 

Test tube	Centrifuge tube	Dropper	Crucible	Watch glass

Burette	Pipette	Graduated cylinder	Funnel	Buchner funnel	Suction flask	Volumetric flask

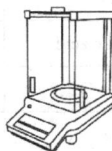 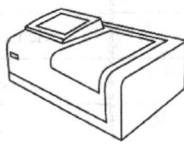  

Top loading balance	Analytical balance	Spectrophotometer	Automatic potentiometric titrator	pH meter

Conductivity meter	Hot plate	Magnetic stirrer hot plate	Water bath	Ultrasonic oscillator

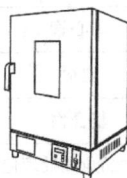  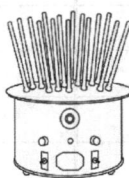 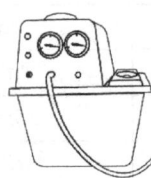 

Oven	Muttle furnace	Airflow dryer	Vacuum pump	Centrifuge

Appx. 2 Vocabulary of Common Professional Words and Phrases
常用专业词汇中英文对照表

英文	中文	英文	中文
absolute error	绝对误差	anhydrate	无水物
absorbance	吸光度	anion	阴离子
absorbance curve	吸收曲线	anion-exchange chromatography	阴离子交换色谱
absorbent paper	吸水纸	argentimetry	银量法
absorptivity	吸收率	ashing	灰化
accuracy	准确度	back titration	返滴定法
acetic acid	乙酸	barium	钡
acetone	丙酮	beaker	烧杯
acid	酸，酸性的	blank test	空白实验
acid-base titration	酸碱滴定	boiling chip	沸石
acidification	酸化	boiling water bath	沸水浴
activated charcoal	活性炭	borate	硼酸盐(根)
activation energy	活化能	borax	硼砂
activity	活度	bromocresol green	溴甲酚绿
activity coefficient	活度系数	Buchner funnel	布氏漏斗
adsorption indicator	吸附指示剂	buffer solution	缓冲溶液
aging	陈化	burette	滴定管
airflow dryer	气流烘干器	calcinate	煅烧
alizarin fluorin blue	茜素氟蓝	calcium	钙
alkali/alkaline	碱/碱性的	calconcarboxylic acid	钙指示剂
aluminum	铝	calibration	校准
aluminum potassium sulfate	硫酸铝钾	carbonate	碳酸盐(根)
ammonia	氨	catalyst	催化剂
ammonia water (solution)	氨水	cation	阳离子
ammonium	铵	cement	水泥
amorphous	无定形的	centrifugation	离心
amphoteric	两性的	centrifuge tube	离心管
amphoteric compound	两性物质	charge	电荷
analyte	分析物	chelate	螯合物
analytical balance	分析天平	chemical spill kit	化学品泄漏处理包
analytical chemistry	分析化学	chloride	氯化物

Continued

英文	中文	英文	中文
chromatographic column	色谱柱	deuterium lamp	氘灯
chromic acid solution	铬酸洗液	dichromate	重铬酸盐(根)
chromium	铬	diethyl ether	乙醚
chromophoric reagent	显色剂	dihydrate	二水合物
citric acid	柠檬酸	dilute	稀释(动词)，稀的
cobalt	钴	dimethylglyoxime	丁二酮肟
colorimetric tube	比色管	diprotic acid	二元酸
colorimetry	比色法	direct titration	直接滴定法
combination electrode	复合电极	dissociation constant	解离常数
complexometric titration	配位滴定	dissolve	溶解
concentrate	浓缩	dithizone	二苯硫腙，二硫腙
concentrated	高浓度的	dodecahydrate	十二水合物
concentration	浓度，浓缩	double salt	复盐
conductivity	电导率	drop	(一)滴
conductivity meter	电导率仪	dropper	滴管
conductometry	电导法	dropper bottle	滴瓶
conjugate base	共轭碱	efflorescent	风化
constant weight	恒重	electrolyte	电解质
continuous variations method	连续变化法(等摩尔系列法)	electromagnetic oven	电磁炉
coordination compound	配合物	electronic balance	电子天平
copper	铜	elementary reaction	基元反应
corrosive	腐蚀性的	emergency shower	紧急喷淋
crucible	坩埚	endpoint	终点
crystal field splitting energy	晶体场分裂能	eriochrome black T (EBT)	铬黑 T
cuvette	比色皿	Erlenmeyer flask	锥形瓶
cyanotype	蓝晒	ethanol	乙醇
decantation	倾析	ethylene diamine tetraacetic acid	乙二胺四乙酸(EDTA)
decompose	分解	evaporating dish	蒸发皿
dehydration	脱水	evaporation	蒸发
deionized water	去离子水	eyewash fountain	洗眼器
desiccator	干燥器	Fajans method	法扬司法
detector	检测器	ferrous ammonium sulfate	硫酸亚铁铵
detergent	洗涤剂	ferrous ion	亚铁离子
determination	测定	filter flask	抽滤瓶

Continued

英文	中文	英文	中文
filter paper	滤纸	indicator electrode	指示电极
filter/filtration	过滤(动词/名词)	indirect titration	间接滴定法
filtrate	滤液	insoluble	难溶的
fire extinguisher	灭火器	interference	干扰
first aid kit	急救药箱	iodide	碘化物/碘离子
first-order reaction	一级反应	iodimetry	直接碘量法
flow chart	流程图	iodine	碘
fluorescein	荧光黄	iodine flask	碘量瓶
fluorescence	荧光	iodometry	间接碘量法
fluoride	氟化物	ionic strength	离子强度
fluorine	氟	ion selective electrode(ISE)	离子选择电极
fume hood	通风橱	iron	铁
funnel	漏斗	lead	铅
gelatinous	胶状的	leakage	渗漏
glass electrode	玻璃电极	ligand	配体
glass rod	玻璃棒	magnetic stirrer	磁力搅拌器
graduated cylinder	量筒	manganese	锰
graduated pipette	分度吸量管(刻度移液管)	mask	掩蔽
gravimetric analysis	重量分析法	measurement error	测定误差
gravity suction	常压过滤	membrane electrode	膜电极
ground	磨口的	meniscus	弯月面
heptahydrate	七水合物	methionine	蛋氨酸(甲硫氨酸)
heteropolyacid salt	杂多酸盐	methyl orange	甲基橙
hexahydrate	六水合物	methyl red	甲基红
hexamethylene tetramine	六次甲基四胺	Mohr method	莫尔法
hot plate	电炉	Mohr's salt	莫尔盐
hydrate	水合物	molar absorptivity	摩尔吸光系数
hydrochloric acid	盐酸	molar mass	摩尔质量
hydrogel	水凝胶	mole ratio method	摩尔比法
hydrogen peroxide	过氧化氢，双氧水	monochromator	单色器
hydrolyze/hydrolysis	水解(动词/名词)	monoclinic	单斜
hydroxide	氢氧化物	mother liquor	母液
hydroxylamine hydrochloride	盐酸羟胺	Muffle furnace	马弗炉
indicator	指示剂	neutralize	中和

Continued

英文	中文	英文	中文
nickel	镍	quantitative chemistry	定量化学
nitrate	硝酸盐(根)	quantitative filter paper	定量滤纸
nitric acid	硝酸	random error	随机误差
oven	烘箱	reaction order	反应级数
oxalate	草酸盐(根)	reaction rate	反应速率
oxalic acid	草酸	recrystallization	重结晶
oxidation state	氧化态	redox titration	氧化还原滴定
oxidize	氧化	reducing agent	还原剂
oxidizing agent	氧化剂	reference electrode	参比电极
permanganate	高锰酸钾	relative average deviation	相对平均偏差
pH meter	酸度计	relative error	相对误差
phenanthroline	邻菲咯啉(邻二氮菲)	replacement titration	置换滴定法
phenolphthalein	酚酞	resin	树脂
phosphate	磷酸盐(根)	rinse	润洗，洗涤
phosphoric acid	磷酸	round off	修约
photosensitivity	光敏性	safety goggles	护目镜
pipette	移液管，移液器	saturated	饱和的
polarity	极性	scientific notation	科学计数法
polyprotic	多元的	scissor	剪刀
potassium	钾	selectivity	选择性
potassium hydrogen phthalate	邻苯二甲酸氢钾	side reaction	副反应
potassium sodium tartrate	酒石酸钾钠	significant figure	有效数字
potassium trioxalatoferrate	三草酸合铁酸钾	silicic acid	硅酸
potentiometric analysis	电位分析法	silver	银
potentiometric titration	电位滴定	sodium	钠
precipitate	沉淀	sodium diphenylamine sulfonate	二苯胺磺酸钠
precipitation titration	沉淀滴定	solubility	溶解度
precision	精密度	solubility product	溶度积
premix	预混	solute	溶质
preparation	制备	spectrophotometer	分光光度计
primary standard	基准物质	spectrophotometry	分光光度法
Prussian blue	普鲁士蓝	spoon	药勺(角匙)
purity	纯度	spot plate	点滴板
qualitative filter paper	定性滤纸	stability constant	稳定常数

Continued

英文	中文	英文	中文
standard addition	标准加入	tip	尖嘴，(移液枪)枪头
standard buffer solution	标准缓冲溶液	titrant	滴定剂
standard curve	标准曲线	titrate/titration	滴定(动词/名词)
standard solution	标准溶液	titration curve	滴定曲线
standardization/standardize	标定(名词/动词)	titration error	滴定误差
starch	淀粉	titration jump	滴定突跃
stepwise titration	分步滴定	titrimetric analysis	滴定分析法
stir	搅拌	transfer	转移
stoichiometric point	化学计量点	transmittance	透光率
stopcock	旋塞，活塞	triethanolamine	三乙醇胺
substitution reaction	取代反应	tungsten	钨
suction filtration	减压过滤(抽滤)	unsaturated	不饱和的
suction flask	抽滤瓶	UV-Vis spectrophotometer	紫外-可见分光光度计
sulfate	硫酸盐(根)	UV-Vis spectrum	紫外-可见光谱
sulfide	硫化物	vacuum pump	真空泵
sulfur	硫	volatile	挥发性的，易挥发的
sulfuric acid	硫酸	Volhard method	福尔哈德法
sulphosalicylic acid	磺基水杨酸	volumetric flask	容量瓶
supernatant	上清液	volumetric pipette	单标线吸量管(胖肚移液管)
supersaturation	过饱和	wash bottle	洗瓶
systematic analysis	系统分析法	watch glass	表面皿
systematic error	系统误差	water-circulation pump	循环水真空泵
tap water	自来水	weak electrolyte	弱电解质
tare	去皮	weighing bottle	称量瓶
tartaric acid	酒石酸	weighing by addition	增量法称量
Teflon stopcock	聚四氟乙烯旋塞	weighing by difference or subtraction	差减法称量，又称减量法称量
test solution	试液	weighing paper	称量纸
test tube	试管	xylenol orange	二甲酚橙
thermometer	温度计	yield	产率
thiosulfate	硫代硫酸盐(根)	zinc	锌
thymolphthalein	百里酚酞		

Appx. 3 Density, Content and Concentration of Common Concentrated Acids and Bases
常用浓酸、浓碱的密度、含量和浓度

Reagent	$\rho /(g \cdot mL^{-1})$ (20°C)	$w/\%$	$c/(mol \cdot L^{-1})$
H_2SO_4	1.84	98	18
HNO_3	1.42	69	16
HCl	1.19	37	12
H_3PO_4	1.7	85	15
$HClO_4$	1.7-1.75	70-72	12
glacial HAc	1.05	99	17
HF	1.13	40	23
$NH_3 \cdot H_2O$	0.88	28	15

Appx. 4 Common Indicators
常用指示剂

1. Simple Acid-base Indicators

Indicator	Color		pK	pT	pH range of color change	Concentration
	Acid form	Alkaline form				
thymol blue(pK_{a1}) 百里酚蓝(第一级解离)	red	yellow	1.6	2.6	1.2-2.8	0.1% (20% ethanol solution)
methyl orange 甲基橙	red	yellow	3.4	4.0	3.1-4.4	0.1% (aqueous solution)
bromocresol green 溴甲酚绿	yellow	blue	4.9	4.4	3.8-5.4	0.1% (20% ethanol solution)
methyl red 甲基红	red	yellow	5.0	5.0	4.4-6.2	0.1% (ethanol solution)
bromothymol blue 溴百里酚蓝	yellow	blue	7.3	7.3	6.0-7.6	0.1% (20% ethanol solution)
thymol blue(pK_{a2}) 百里酚蓝(第二级解离)	yellow	blue	8.9	9.0	8.0-9.6	0.1% (20% ethanol solution)
phenolphthalein 酚酞	colorless	red	9.1	9.0	8.2-9.8	0.1% (ethanol solution)
thymolphthalein 百里酚酞	colorless	blue	10.0	10	9.4-10.6	0.1% (ethanol solution)

2. Mixed Acid-base Indicators

(all the concentrations are 0.1%)

Indicator 1 (solvent)	Indicator 2 (solvent)	Color			pT	pH range	$V : V$
		Acid color	Alkaline color	Endpoint color			
cresol yellow (ethanol) 甲酚黄(乙醇)	methylene blue (ethanol) 亚甲基蓝(乙醇)	violet	green		3.25	3.2-3.4	1 : 1

Continued

Indicator 1 (solvent)	Indicator 2 (solvent)	Color			pT	pH range	$V : V$
		Acid color	Alkaline color	Endpoint color			
methyl orange (water) 甲基橙(水)	thymol blue sodium (water) 百里酚蓝钠(水)	violet	yellow-green	gray	4.1	3.1-4.4	1 : 1
bromocresol green (ethanol) 溴甲酚绿(乙醇)	methyl red (ethanol) 甲基红(乙醇)	wine-red	green		5.1		3 : 1
neutral red (ethanol) 中性红(乙醇)	methylene blue (ethanol) 亚甲基蓝(乙醇)	violet	green	violet	7.0		1 : 1
cresol red (water) 甲酚红(水)	thymol blue sodium (water) 百里酚蓝钠(水)	yellow	violet	pinkish purple	8.3	8.2-8.4	1 : 3
phenolphthalein (ethanol) 酚酞(乙醇)	thymol (ethanol) 百里酚(乙醇)	colorless	violet	purple red	9.9	9.6-10.0	1 : 1

3. Metal Ion Indicators

Indicator	Color		pH range	Concentration	Application
	In	MIn			
eriochrome black T 铬黑 T (EBT)	blue	red	7-10	1% EBT(1 g EBT + 75 mL trolamine + 25 mL ethanol)	pH = 10: Mg^{2+}, Zn^{2+}, Cd^{2+}, Pb^{2+}, Hg^{2+}, In^{3+}
xylenol orange 二甲酚橙 (XO)	yellow	red	< 6	0.2% (aqueous solution)	pH = 1-2: Bi^{3+} pH = 5-6: Cd^{2+}, Co^{2+}, Cu^{2+}, Pb^{2+}
1-(2-pyridylazo)-2-naphthol 1-(2-吡啶偶氮)-2-萘酚(PAN)	yellow	red	2-12	0.3% (ethanol solution)	pH = 5-6: Cd^{2+}, Ni^{2+}, Cu^{2+}, Pb^{2+}
calconcarboxylic acid 钙指示剂	blue	red	10-13	calconcarboxylic acid : NaCl = 1 : 100 ($m : m$)	pH = 12-13: Ca^{2+}
acid chrome blue K 酸性铬蓝 K	blue	red	8-13	0.1% (ethanol solution)	pH = 10: Mg^{2+}, Zn^{2+}
K-B indicator K-B 指示剂	blue	red	8-13	1 g acid chrome blue K + 0.4 g naphthol green B + 40 g KCl	pH = 10: Mg^{2+}, Zn^{2+}, Ca^{2+}
sulphosalicylic acid 磺基水杨酸	colorless	red	< 8	1%-2% (aqueous solution)	pH = 1-1.5: Fe^{3+}

4. Redox Indicators

Indicator	Color		$E^{\ominus} / V$ $[H^+]=1 \ mol \cdot L^{-1}$	Concentration
	Reduction state	Oxidation state		
methylene blue 亚甲基蓝	colorless	blue	0.52	1% (aqueous solution)
diphenylamine 二苯胺 [a]	colorless	purple	0.76	0.1% (concentrated H_2SO_4 solution)
sodium diphenylamine sulfonate 二苯胺磺酸钠	colorless	reddish-violet	0.85	0.2% (aqueous solution)
ferroin 邻二氮菲亚铁 [b]	red	light blue	1.06	0.025 $mol \cdot L^{-1}$ (aqueous solution: 1.485 g 1, 10- phenanthroline and 0.695 g $FeSO_4 \cdot 7H_2O$, dissolve in water, dilute to 100 mL)
nitroferroin 硝基邻二氮菲亚铁	red	light blue	1.25	0.025 $mol \cdot L^{-1}$ (aqueous solution)

Note: a) Interference by W; b) Interference by strong acid salts of Cd, Zn, Cu, Ni, Co.

5. Adsorption Indicators

Indicator	Color change	Titration conditions	Measured ion	Titrant
fluorescein 荧光黄	yellowish green → red	pH 7-10	Cl^-、Br^-、I^-、SCN^-	Ag^+
dichloro fluorescein 二氯荧光黄	yellowish green → red	pH 4-10	Cl^-、Br^-、I^-	Ag^+
eosin 曙红	red-orange → red-purple	pH 2-10	Br^-	Ag^+
rhodamine 6G 罗丹明 6G	orange red → reddish purple	acidic medium	Ag^+	Br^-
alizarin red 茜素红	yellow → red	acidic medium	SCN^-	Ag^+

Appx. 5　Color of Common Ions and Inorganic Compounds
常见离子和无机化合物的颜色

1. Color of Common ions

Ion	Color	Ion	Color	Ion	Color	Ion	Color
Ac^-	colorless	Cr^{3+}	green	Mg^{2+}	colorless	Pb^{2+}	colorless
Ag^+	colorless	CrO_4^{2-}	yellow	Mn^{2+}	light pink	PO_4^{3-}	colorless
Al^{3+}	colorless	$Cr_2O_7^{2-}$	orange	MnO_4^-	reddish-violet	SCN^-	colorless
Ba^{2+}	colorless	Cu^{2+}	blue	MnO_4^{2-}	green	SiO_3^{2-}	colorless
Bi^{3+}	colorless	Fe^{2+}	light green	MoO_4^{2-}	colorless	Sn^{2+}	colorless
Ca^{2+}	colorless	Fe^{3+}	yellow	Na^+	colorless	SO_3^{2-}	colorless
Cl^-	colorless	$[Fe(CN)_6]^{3-}$	yellow	NH_4^+	colorless	SiO_4^{2-}	colorless
Co^{2+}	pink	$[Fe(CN)_6]^{4-}$	yellow	Ni^{2+}	green	$S_2O_3^{2-}$	colorless
CO_3^{2-}	colorless	H^+	colorless	NO_2^-	colorless	SO_4^{2-}	colorless
$C_2O_4^{2-}$	colorless	I_3^-	brownish yellow	NO_3^-	colorless	Sr^{2+}	colorless
Cr^{2+}	blue	K^+	colorless	OH^-	colorless	Zn^{2+}	colorless

2. Color of Common Inorganic Compounds

White or colorless: $AgCl$, $AgIO_3$, $AgNO_3$, $AgSCN$, $AlCl_3 \cdot 6H_2O$, $Al(NO_3)_3 \cdot 9H_2O$, Al_2O_3, $Al(OH)_3$, $Al_2(SO_4)_3$, $BaCl_2$, $BaCO_3$, $BaSO_4$, $Bi(NO_3)_2$, $CaCl_2$, $CaCl_2 \cdot 6H_2O$, $CaCO_3$, CaC_2O_4, CaO, $Ca(OH)_2$, $CaSO_4$, $CuSO_4$, EDTA-2Na, HAc, H_3BO_3, HCl, $HClO_4$, $H_2C_2O_4$, HF, HIO_3, HNO_3, H_2O_2, H_3PO_4, H_2S, H_2SO_4, $KAl(SO_4)_2 \cdot 12H_2O$, KBr, KCl, KCN, K_2CO_3, $K_2C_2O_4 \cdot H_2O$, KF, KIO_3, KNO_3, KSCN, K_2SO_4, $KHC_8H_4O_4$, KH_2PO_4, KI, KOH, K_3PO_4, K_2SO_4, $K_2S_2O_8$, $K_2WO_4 \cdot 2H_2O$, $LiAlH_4$, $MgCl_2$, $Mg(NO_3)_2 \cdot 6H_2O$, MgO, $Mg(OH)_2$, $MgSO_4 \cdot 7H_2O$, NaAc, NaCl, Na_2CO_3, $Na_2C_2O_4$, NaF,

$NaHCO_3$, $Na_2HPO_4 \cdot 12H_2O$, $NaH_2PO_4 \cdot 2H_2O$, $NaNO_3$, $NaOH$, $Na_3PO_4 \cdot 12H_2O$, $NaSCN$, Na_2SiO_3, Na_2SO_3, Na_2SO_4, $Na_2S_2O_3$, $Na_2WO_4 \cdot 2H_2O$, NH_4Cl, $(NH_4)_2CO_3$, $(NH_4)_2C_2O_4 \cdot H_2O$, NH_4F, NH_4HCO_3, $(NH_4)_6Mo_7O_{24} \cdot 4H_2O$, NH_4NO_3, NH_4SCN, $(NH_4)_2SO_4$, $PbCO_3$, $PbCl_2$, $Pb(NO_3)_2$, $Pb(OH)_2$, $PbSO_4$, SiO_2, $SnCl_2$, SO_2, SO_3, $SrCl_2 \cdot 6H_2O$, $Sr(NO_3)_2$, TiO_2, $ZnCl_2$, $ZnCO_3$, $Zn(NO_3)_2 \cdot 6H_2O$, ZnO, $Zn(OH)_2$, $ZnSO_4 \cdot 7H_2O$

Formula	Color	Formula	Color	Formula	Color
Ag_2O	brown or black	CuO	black	MnC_2O_4	light red
Ag_2S	black	$Cu(OH)_2$	blue	MnO_2	black
$AgBr$	light yellow	$CuSO_4 \cdot 5H_2O$	blue	$Mn(NO_3)_2$	light red
AgI	yellow	$FeCl_3 \cdot 6H_2O$	yellowish brown	$MnSO_4 \cdot H_2O$	light red
$Ca(H_2PO_4)_2 \cdot H_2O$	colorless or yellowish	$Fe(NO_3)_3 \cdot 9H_2O$	light purple or colorless	$NaNO_2$	light yellow or white
$Co(Ac)_2 \cdot 4H_2O$	dark red	Fe_2O_3	reddish brown	$NH_4Fe(SO_4)_2 \cdot 12H_2O$	gray-purple
$CoCl_2$	blue	$Fe(OH)_3$	reddish brown	$(NH_4)_2Fe(SO_4)_2 \cdot 6H_2O$	light green or blue-green
$CoCl_2 \cdot 6H_2O$	pink	$FeSO_4 \cdot 7H_2O$	green	$NiCl_2 \cdot 6H_2O$	green
$Co(NO_3)_2 \cdot 6H_2O$	dark red	K_2CrO_4	yellow	$Ni(NO_3)_2 \cdot 6H_2O$	green
$CoSO_4 \cdot 7H_2O$	red	$K_2Cr_2O_7$	red-orange	$NiSO_4 \cdot 6H_2O$	green
$CrCl_3 \cdot 6H_2O$	dark green	$K_3[Fe(CN)_6]$	red	$Ni(OH)_2$	green
$Cr(NO_3)_3 \cdot 9H_2O$	dark purple	$K_4[Fe(CN)_6]$	yellow	NO_2	reddish brown
$Cu(Ac)_2 \cdot H_2O$	dark green	$K_3[Fe(C_2O_4)_3] \cdot 3H_2O$	green	PbO	yellow
$Cu(NO_3)_2 \cdot 3H_2O$	blue	$KMnO_4$	purple	$TiCl_4$	light yellow or colorless

Appx. 6   Bilingual Videos Embedded in the Manual
本书插入双语视频清单

No.	Bilingual videos	Chapter	Page
1	视频 1-1 化学实验室安全 Safety in the Chemistry Laboratory	1	6
2	视频 1-2 课程要求 Requirements for the Course	1	8
3	视频 1-3 实验记录本、预习报告和实验报告指南 Guidelines for Lab Notebook, Pre-lab, and Post-lab Report	1	12
4	视频 1-4 滴定分析法概述 Introduction to Titrimetric Analysis	1	16
5	视频 1-5 有效数字 Significant Figures	1	21
6	视频 1-6 化学分析中的误差 Errors in Chemical Analysis	1	21
7	视频 2-1 容量瓶的基本操作 Basic Operations of Volumetric Flask	2	27
8	视频 2-2 滴定管的基本操作 Basic Operations of Burette	2	29
9	视频 2-3 移液管的基本操作 Basic Operations of Pipette	2	31
10	视频 2-4 移液器的基本操作 Basic Operations of Automatic Pipette	2	32
11	视频 2-5 减压过滤的基本操作 Basic Operations of Suction Filtration	2	36
12	视频 2-6 分析天平的基本操作 Basic Operations of Analytical Balance	2	39

Continued

No.	Bilingual videos	Chapter	Page
13	视频 2-7　分光光度法概述　Introduction to Spectrophotometry	2	42
14	视频 2-8　分光光度计的使用　How to Use a Spectrophotometer	2	42
15	视频 2-9　酸度计的使用　How to Use a pH Meter	2	44
16	视频 3-1　常见阳离子的分离和鉴定(讲解) Separation and Identification of Common Cations	3	53
17	视频 3-2　常见阳离子的单个鉴定(实验) Individual Confirmatory Tests for Common Cations	3	53
18	视频 3-3　已知阳离子混合液的分离和鉴定(实验) Separation and Identification of Known Mixture of Cations	3	53
19	视频 3-4　离心、沉淀洗涤和滴瓶使用 Centrifugation, Washing of Precipitates, Using of Dropper Bottles	3	53
20	视频 3-5　常见阴离子的分离和鉴定(讲解) Separation and Identification of Common Anions	3	60
21	视频 3-6　常见阴离子的分离和鉴定(实验) Separation and Identification of Common Anions	3	60
22	视频 3-7　速率常数和活化能的测定(讲解) Determination of Rate Constant and Activation Energy	3	66
23	视频 3-8　速率常数和活化能的测定(实验) Determination of Rate Constant and Activation Energy	3	66
24	视频 3-9　用计算机处理数据和作图　Computer Processing Data and Plotting	3	66
25	视频 3-10　明矾的制备及其大晶体培养(讲解) Preparation of Alum and Large Crystal Cultivation	3	71
26	视频 3-11　明矾的制备及其大晶体培养(实验) Preparation of Alum and Large Crystal Cultivation	3	71
27	视频 4-1　酸碱滴定法概述　Introduction to Acid-base Titration	4	74
28	视频 4-2　NaOH 溶液的配制与标定　Preparation and Standardization of NaOH Solution	4	77
29	视频 4-3　HCl 溶液的配制与标定　Preparation and Standardization of HCl Solution	4	77
30	视频 4-4　氧化还原滴定法概述　Introduction to Redox Titration	4	78
31	视频 4-5　$KMnO_4$ 溶液的配制与标定 Preparation and Standardization of $KMnO_4$ Solution	4	80
32	视频 4-6　$Na_2S_2O_3$ 溶液的配制与标定　Preparation and Standardization of $Na_2S_2O_3$ Solution	4	84
33	视频 4-7　配位滴定法概述　Introduction to Complexation Titration	4	86
34	视频 4-8　EDTA 溶液的配制与标定　Preparation and Standardization of EDTA Solution	4	88
35	视频 4-9　沉淀滴定法概述　Introduction to Precipitation Titration	4	92
36	视频 4-10　酸碱滴定练习　Practice of Acid-base Titration	4	94
37	视频 4-11　未知酸分子量的测定　Determination of the Molecular Weight Of An Unknown Acid	4	97
38	视频 5-1　氟离子选择电极测定茶叶中氟含量(讲解) Determination of Fluoride Content in Teas Using Fluoride Ion Selective Electrode	5	104
39	视频 5-2　氟离子选择电极测定茶叶中氟含量(实验) Determination of Fluoride Content in Teas Using Fluoride Ion Selective Electrode	5	104
40	视频 5-3　硫酸亚铁铵的制备及表征(讲解) Preparation and Characterization of Ferrous Ammonium Sulfate	5	107
41	视频 5-4　硫酸亚铁铵的制备及表征(实验) Preparation and Characterization of Ferrous Ammonium Sulfate	5	107
42	视频 5-5　蛋氨酸铜的制备及表征(讲解) Preparation and Characterization of Copper Methionine	5	111

Continued

No.	Bilingual videos	Chapter	Page
43	视频 5-6　蛋氨酸铜的制备及表征(实验) Preparation and Characterization of Copper Methionine	5	111
44	视频 5-7　三水合三草酸合铁(III)酸钾的制备及光敏性测试(讲解) Preparation and Photosensitivity Test of Potassium Trioxalatoferrate(III) Trihydrate	5	115
45	视频 5-8　三水合三草酸合铁(III)酸钾的制备及光敏性测试(实验) Preparation and Photosensitivity Test of Potassium Trioxalatoferrate(III) Trihydrate	5	115
46	视频 5-9　三草酸合铁(III)配离子电荷数的测定(讲解) Determination of the Charge Number of Trioxalatoferrate (III) Complex Ion	5	118
47	视频 5-10　三草酸合铁(III)配离子电荷数的测定(实验) Determination of the Charge Number of Trioxalatoferrate (III) Complex Ion	5	118
48	视频 5-11　分光光度法测定条件优化(讲解) Optimization of Spectrophotometric Determination Conditions	5	125
49	视频 5-12　分光光度法测铁(讲解) Spectrophotometric Determination of Iron	5	125
50	视频 5-13　分光光度法测铁(实验) Spectrophotometric Determination of Iron	5	125
51	视频 5-14　三种钴氨配合物的制备及表征(讲解) Synthesis and Characterization of Three Kinds of Cobalt-ammine Coordination Compounds	5	131
52	视频 5-15　三种钴氨配合物的制备(实验) Synthesis of Three Kinds of Cobalt-ammine Coordination Compounds	5	131
53	视频 5-16　钴氨配合物的钴含量测定(实验) Determination of Cobalt Content of Cobalt-ammine Coordination Compounds	5	131
54	视频 5-17　重量分析法概述 Introduction to Gravimetric Analysis	5	139
55	视频 5-18　水泥组分测定(讲解) Determination of Cement Composition	5	139
56	视频 5-19　水泥样品处理和 SiO_2 含量测定(实验) Treatment of Cement Sample and Determination of SiO_2 Content	5	139
57	视频 5-20　水泥样品中铁、铝、钙、镁含量测定(实验) Determination of Fe Al Ca Mg Contents in Cement Sample	5	139
58	视频 5-21　三种磷酸钠盐水合物的制备及表征(讲解) Preparation and Characterization of Three Kinds of Sodium Phosphate Hydrates	5	144
59	视频 5-22　三种磷酸钠盐水合物的制备(实验) Preparation of Three Kinds of Sodium Phosphate Hydrates	5	144
60	视频 5-23　自动电位滴定仪的使用 Operations of Automatic Potentiometric Titrator	5	144
61	视频 5-24　杂多酸盐的制备及其速率常数测定(讲解) Preparation of Heteropolyacid Salts and Determination of the Rate Constant	5	150
62	视频 5-25　杂多酸盐的制备及其速率常数测定(实验) Preparation of Heteropolyacid Salts and Determination of the Rate Constant	5	150

Appx. 7　Atomic Weight of Elements
元素的原子量

(蔡吉清　曾秀琼编写)